DOCUMENTATION BASICS

That Support Good Manufacturing Practices

Carol DeSain

Advanstar Communications
Cleveland, Ohio

*For Charmaine Vercimak, Georgiann Keyport,
Diane Krizek and all the other young people
I have worked with in this industry who have
kept me honest, allowed me to grow and
forced me to communicate by ALWAYS
questioning what I tell them!*

—*CD*

Copyright © 1993 Carol DeSain

All rights reserved. No part of this book may be reproduced or used in any form or by any means—graphic, electronic, or mechanical, including photocopying, recording, taping, or information storage and retrieval systems—without written permission of the publisher.

Printed in the United States of America

10 9 8 7 6 5 4 3 2

Library of Congress Cataloging-in-Publication Data

DeSain, Carol.
    Documentation basics that support good manufacturing practices/by Carol DeSain.
      p.   cm.
    Includes bibliographical references.
    ISBN 0-943330-30-0 : $75.00
    1. Pharmaceutical industry—Quality control—Documentation.
2. Manufactures—Quality control—Documentation. I. Title.
HD9665.5.D4   1993
615.1'068'5—dc20   92-31545
                                                                                      CIP

Published by Advanstar Communications

Advanstar Communications produces technical books serving a wide range of industries. The company also publishes a complete family of magazines and technical journals, including *Applied Clinical Trials, BioPharm, CADalyst, GPS World, Geo Info Systems, LC•GC, Managed Health Care News, Pharmaceutical Executive, Pharmaceutical Technology,* and *Spectroscopy.*

For additional information on any magazines or a complete catalog of Advanstar books, please write to Advanstar Marketing Services, 7500 Old Oak Boulevard, Cleveland, OH 44130.

Design by Sandra J. Harner, Laing Communications Inc.

Contents

		Introduction	v
1	■	*Part Numbers*	1
2	■	*Part Number Specifications*	7
3	■	*Lot Numbers*	17
4	■	*Standard Operating Procedures*	27
5	■	*Master Production Batch Records*	35
6	■	*Equipment Installation and Identification*	47
7	■	*Equipment Monitoring, Repair and Preventive Maintenance*	55
8	■	*Designing GMP and Facility Qualification Master Protocols*	61
9	■	*Master Validation and Equipment Validation Protocols*	67
10	■	*Master Method Validation Protocols*	75
11	■	*Process Validation Protocols*	79
12	■	*Good Business Practices*	85
		References	87

Introduction

A DOCUMENTATION SYSTEM is the currency of a corporation. It is the "level playing field" on which employees communicate with one another and, as a result, its proper design and implementation directly influence how effectively information is shared and problems are solved.

Documentation is integral to a company's internal and external interactions. Consequently, the documentation system must be a proactive vehicle of communication, designed primarily to be useful. It must also be flexible enough to grow and change with the company.

The documentation system described in this text is designed to support Good Manufacturing Practices (GMP) in a medical manufacturing environment. However, the usefulness of the example can be extended to other areas of the corporation, i.e., development, clinicals, marketing, finance, as well as to many other unrelated, nonmedical industries. The principles and the decision making inherent in documentation system design remain the same, no matter what the product or business.

As in any system, there are basic components which inter-relate to serve a common system function. In a documentation system these components are descriptive documents, data collection documents, numbering systems and data files.

Descriptive documents describe how to perform certain tasks, such as standard operating procedures, protocols, specifications and master production records.

Data collection documents facilitate the timely and accurate documentation of tasks and events, such as forms, reports, production batch records and logbooks.

Numbering systems serve to account for and track information and documents, such as part numbers, lot numbers, equipment numbers, form numbers, SOP numbers, receiving codes, etc.

Data files serve to organize the data into useful categories of concern for review and to support accountability and traceability requirements. Data files include specification files, equipment history files, product files and facility qualification files.

The primary function of a documentation system in a GMP operation is to establish, monitor and document quality. The descriptive documents define and establish the quality of the raw materials, environment, production process and final product. The data collection documents confirm that the materials, environment, production process or product routinely meet the established quality characteristics. The numbering systems serve to control and track the use of descriptive documents and data collection documents. Data files organize the data into useful categories to facilitate review and retrieval.

This book describes the creation, use and control of the descriptive documents, data collection documents, numbering systems and data files that are appropriate for use in an industry subject to Good Manufacturing Practices. There are two ways to present such a documentation system. The first is a top-down approach that presents the general commitments of the system and then describes how the system is implemented. The second is a bottom-up approach that describes the construction of the most fundamental building blocks of the system first, then builds on this structure.

Ultimately these different viewpoints reflect the fact that there are two types of users: Manufacturing personnel who interact with the "nuts and bolts" of the system daily and need to know what to do and how to do it, and administrative personnel (regulatory, development, finance, marketing) who must understand and interact with the system by reference. This text was written as a guideline for the individuals who must design the systems and then work with them routinely.

The descriptive documents presented in this book and forthcoming, associated workbooks are designed to serve two purposes: to direct task-specific events and to educate the reader about the event in a manner that supports responsible decision making. For example, a procedure about part number assignment and specification writing describes not only how to assign part numbers and write specifications, it also describes the purpose of part numbers and specifications and provides guidance on the decision making associated with their assignment and use. When a document is written both to inform and to educate, it must meet the needs of the line worker *and* the administrator.

How one makes a decision in a GMP facility must be controlled as much as any measurable product characteristic. Inconsistent decision making is as much of a threat to product quality as uncontrolled change. For these reasons, this text describes how to create the components of a documentation system and presents guidelines for the decision making required for an effective system of communication.

Documenting the rationale for responsible decision making is integral to the philosophy of Good Manufacturing Practices. This seems obvious when one considers the need to document the rationale for changing a specification or changing a manufacturing process; it is not so obvious when one considers the need to document the rationale behind the design of a documentation system. Without the rationale, without guidelines for decision making, the system would be vulnerable to major inconsistencies, and the work would proceed very slowly. Either possibility costs time and money and is a source of tremendous aggravation.

As has been emphasized so dramatically in the International Standards Organization (ISO) 9000 guidelines, the documentation of a quality assurance system may soon be a requirement for "doing business" in the international marketplace. As a result it is not only instructive but imperative to understand how to design these systems so that they support routine operations.

A well-designed documentation system ensures that quality standards are met routinely, minimizes the potential for error, reduces downtime when deviations or failures occur by providing immediate access to well-organized data, and serves as a consistent training tool for line workers and administrators alike. A poorly designed system is a burden to all.

In a medical manufacturing environment, a sound, sensible and simple system of documentation is the foundation of Good Manufacturing Practices. Although that statement may seem obvious, and the creation of the documentation system may seem straightforward, anyone who has attempted the task realizes very quickly that the best of intentions can be quickly sabotaged in several ways. Not all departments that will use the system understand or agree upon the basic purpose of each numbering system or each type of document. When one numbering system is designed to achieve two purposes instead of just one, the integral nature of the entire documentation system can be complicated unnecessarily or lost entirely. Numbers and documents may be created by people who do not routinely use them, and there are bound to be people who believe that "numbers are numbers" and that a system that worked well in an engineering environment or a medical device company will be "just fine" for biologics.

During the development of a product there comes a time when a small R&D company decides that GMP operations are required. The company creates a quality control manager by expanding the job description of a current research technician or scientist and assigns the task of GMPs. Anyone who has had this experience knows how overwhelming it can be because no reference sources explain how to establish such operations. The Code of Federal Regulations, the Food and Drug Administration's Guidelines, and Points to Consider are vague. Workshops, seminars, and conferences can be intimidating. You may finally conclude that designing these systems must be such a simple task that everyone already knows how to do it—everyone but you! That's not true.

A simple system is always the most difficult to design because there seems to be unending pressure to add "just one more thing." Bosses and other allied critics offer not-so-constructive comments on components of the system without taking the time to understand and appreciate the big picture. When tentative about one's expertise to begin with, one is vulnerable to these opinions. Clearly, anyone who claims it's easy has never successfully designed a GMP-based documentation system!

Documentation Basics will present the major components of a GMP documentation system, give examples of design, format and content, and then explain how these components interact.

1

Part Numbers

A PART NUMBERING SYSTEM is a basic building block of GMP documentation. A part number is a simple code used to identify one of a variety of items while a part number specification, discussed in the next chapter, defines its identity, describes the item in detail and can include a way to test the quality of that item.

Traditional manufacturing industries—like automobile and refrigerator manufacturers—use part numbers to facilitate purchasing events, to track the flow of materials throughout a facility and to support inventory-control practices. Part numbers in a GMP operation share these functions but have a different primary purpose.

In a GMP operation, the quality of an item differentiates it from similar items. For example, the part number for a 500-gram bottle of sodium chloride, ACS grade, will have a different part number than a 500-gram bottle of sodium chloride, USP grade.

In a non-GMP operation, the configurations of items alone might result in different part numbers. For example, a 2.5-kilogram bottle of sodium chloride, ACS, might have a part number different from that of a 500-gram bottle of sodium chloride, ACS. In a GMP industry, these two items would have the same part number because the salt in both containers is of the same quality. Part numbers in GMP operations are used primarily to control quality. This quality control is fundamental to GMP operations.

Part numbers are used to identify items purchased from outside vendors and can also be used to identify items formulated or produced within the facility (solutions, components and product). Ultimately, part numbers are used because they are convenient. Using numbers rather than long narrative descriptions permits one to quickly identify items as the same or different from one another. In daily use a technician is more easily able to write down a short number than a long description.

Which Items to Number

Critical items require part numbers. These are items used in the manufacture of final product—they are part of a product or come in direct contact with the final product during processing. They may be purchased from outside vendors or prepared in-house. Vials, stoppers and seals, for example, are critical components of a traditional pharmaceutical product and are usually purchased from an outside vendor. Similarly, cells, media, growth factors, bioreactors and gas-exchange cartridges are critical components in a hollow-fiber-based cell culturing system.

Any item whose failure could directly and adversely affect the quality, efficacy or safety of the final product needs a part number. Some items that never come in direct contact with the product can be very important in establishing its quality. For example, reagent standards, analytical column resins or a biological substrate for a product-potency assay may be critical to the ultimate quality and integrity of the product.

Any item that requires control to ensure its quality requires a part number. This includes materials subject to vendor expiration dating, materials with special handling requirements and materials known to change with time.

Any item whose control can provide convenience or ultimately assist in troubleshooting analytical assays or production processes needs a part number. In many quality control laboratories all reagents are assigned part numbers. When performing assays, whether for research or to support production of a product, it is far more convenient to cite a part number when preparing a solution than to write it out longhand. Also, the ability to track reagent use is invaluable when troubleshooting an assay.

Items Not Requiring Part Numbers

In a small company it is unrealistic and unnecessary to assign part numbers to supplies such as aluminum foil, laboratory notebooks and general glassware. A time may come as the company grows when a purchasing agent insists on part numbers for everything. If that day arrives, then these part numbers should be visibly different from the part numbers already assigned, even if they are merely a separate category of numbers, as will be discussed later. The supply part numbers, if and when they are created, must be different because the supplies will have no part number specifications, and they will not pass through a quality control inspection before being released into the facility.

How to Assign Part Numbers

Part numbers should be assigned by the quality control (QC) or quality assurance (QA) department. There should be only one part number for each unique item.

It is generally helpful to design a part number request form that describes the item in detail, designates vendors and catalog numbers, describes how the item will be used and suggests any critical features or parameters to be checked when the item arrives. The importance of this form will become apparent later when part number specifications are discussed.

The QA manager who receives such a request assigns a number. Because part numbers are requested just prior to ordering an item for the first time (the purchase order request form usually asks for a part number), providing the part number request form to QA as early as possible will expedite release of the item when it arrives on the dock.

Assigned and yet-to-be-assigned part numbers can be kept in a notebook with manual entries. At a convenient time, these entries can be transferred to an electronic database to facilitate their use.

How to Name a Part

The narrative description of a part number is important to the maintenance of accurate and effective lists. Each part number must have a unique name that allows one to find the item quickly and determine how or why this item is different from other items on the list. For example, a 10-inch sterilizing cartridge filter will have a different part number than will a 20-inch sterilizing filter. If one were looking at the part number list for these items, however, one would probably begin by looking under "F" for filter. Therefore, part number names should be "Filter, 0.2 µm, cellulose, 10-inch cartridge" and "Filter, 0.2 µm, cellulose, 20-inch cartridge."

Include only those critical qualifiers that distinguish this item from other items on the list. Appropriate qualifiers include chemical grades. For example "Sodium chloride, USP" would have a different part number than "Sodium chloride, ACS." Brand names should appear in the part number name only if it is determined that there is no acceptable alternative vendor for the item.

Part Number Categories

Part numbers can be simply random numbers. It is convenient, however, to organize the part numbers into identifying categories. For example, if a four-digit part number is sufficient, then it may make sense to develop categories for the 1000-series, 2000-series, and so on.

This categorization is useful as long as the categories serve only to extend the purpose of identification. Do not assign part number categories to individual departments, such as QC or production. A part number system must ultimately ensure Good Manufacturing Practices for the entire facility, and its design and use should reflect that. Do not assign categories based on the price of an item—for example, items costing more than $2000. An accounting-driven system contradicts the primary purpose of a GMP documentation system.

Categories should reflect either different types of items or different levels of quality concern. For example, in a contract vial-filling facility, it may be convenient to have vials in the 3000-series, stoppers in the 4000-series and seals in the 5000-series. Critical inventory-controlled items might be placed in a different category than noncritical items.

The usefulness of this last example becomes apparent when one tries to interface the material handling and inventory control system with the part numbering system. Clearly, in a small company inventory control, which accommodates the requirements of the IRS and the SEC, need apply only to those items that are critical and/or costly to operations. An easily recognizable category of part numbers can sometimes facilitate this identification.

This is not a perfect system, however. There are always expensive items that are not critical to product quality yet are important to an accountant. Accounting requirements must be accommodated, but they can be accommodated by extending inventory control to certain noncritical items rather than compromising the primary purpose of the part numbering system, which is to define and control quality.

Also, accountants often exert pressure to use the part numbering system to distinguish between chemicals and components used for research and those used in manufacturing. This is asking too much from a part numbering system, especially in a facility that ultimately uses the same supplies and reagents for both tasks. This division or assessment should be made another way.

Cell Line Part Numbers

Cell lines or cell substrates are critical raw materials used in the production and testing of final product; as a result, they must be completely identified and their quality controlled. Assigning a part number to each unique cell line will facilitate this process. Cell lines should be assigned their own unique numbering category, discussed above, because their identifying characteristics and quality parameters differ markedly from other items in the part numbering system.

Each unique cell line must have a unique part number. Differences between cell lines can be subtle and sometimes unknown. Clearly, if one has two apparently identical cell lines from different sources or customers, the cell lines should receive different part numbers.

Identifying cell lines presents some special difficulties because passage level is a critical parameter, and a master cell bank (MCB) sample must be distinguished from a working cell bank (WCB) sample. This issue will be discussed in the chapter on lot numbers.

Process Intermediate Part Numbers

There are likely to be points during the processing or purification of a product at which intermediate products are produced and their quality is evaluated before processing proceeds. There are a couple of ways to control this. If the processing intermediate will be stored for some time before processing continues, then the intermediate should be identified with a part number (a critical-chemical part number); the criteria for release of that intermediate can be described in its part number specification. Specification content will be described in the next chapter.

If, however, the processing intermediate is held for a short time (overnight or until QC analysis is complete), then the testing requirements and the acceptance criteria should appear directly on the batch record.

Part Numbers Never Reassigned

Once a part number is assigned to an item and that item has been used in the facility—meaning that the part number has appeared on GMP documentation—then that part number should never be reassigned. Clearly, a number can be retired but not reassigned. Records are open to Food and Drug Administration review and legal due diligence procedures for many years. It is sensible, therefore, to retire rather than compromise the outcome of legal audits.

Operational Suggestions

A part numbering system should be supported by two lists: A numerical list of part numbers (by category) and part names, and an alphabetical list of parts and part numbers. All departments in the manufacturing facility should have access to updated lists.

Part number request sheets should be developed to facilitate part-number assignment and begin the process of part number specification writing. In addition, a standard operating procedure (SOP) should be written that describes how the system operates, who assigns numbers, how they are controlled, how to complete a part number request sheet and how QA categorizes part numbers.

As this text continues, part numbers based on the following numbering system will be used as examples in other documents. (Figure 1.1)

A Sample Numbering System

1000–1999	Critical chemicals and intermediate product*
2000–2999	Critical components*
3000–3999	Cell lines*
4000–4999	Final product*
5000–5999	Printed materials and labels*
6000–6999	Noncritical chemicals
7000–7999	Noncritical components
8000–9999	Solutions prepared in-house

*Inventory-controlled items

(Figure 1.1)

2

Part Number Specifications

A SPECIFICATION DOCUMENT is the primary source of information for an item. It is a form that documents the description of each part numbered item, contains purchasing information, chemical formulas, dimensions, sampling information, handling precautions, storage conditions and, when required, testing methods and acceptance criteria.

When a part numbered item is received from a vendor, it is assigned a receiving code, labeled/identified with its part number and receiving code and placed into a quarantined area. QC must inspect and/or test this item before it can be released into a GMP facility. The specification, therefore, identifies the item and describes how it is tested by QC to confirm its identity and quality before it is acceptable for use in the manufacturing facility.

Each page of a specification should contain the name of the company, a title or narrative description of the item, the part number of the item, the edition number of the specification (discussed below), pagination (page 1 of 3, for example) and approval signatures. In addition, each specification should contain a list of approved vendors for that item with catalog numbers, an expiration date for the item, sample size (when testing is required), file sample size (when required for critical items), handling precautions and testing or acceptance criteria.

Each category of part numbered items can have its own specification form to facilitate documentation of its characteristics. The following examples are guides for designing category-specific specifications forms.

A **chemical specification form** can contain a physical appearance description of the chemical, chemical grade, empirical formula and formula weight. (Figure 2.1)

A **component specification form** can contain sections on material composition, size, dimensions and color.

A **cell-line specification form** can indicate whether the cell line is suspension- or anchorage-dependent and detail cell-line history, vectors, markers, passage level limitations, isoenzyme analysis, etc.

A **printed materials** or **labeling specification form** can contain a color chart (when appropriate) and an actual approved master copy of the printed item to use for comparison during release work. Approval signatures for a printed item or label should be signed on the item itself, if possible.

A **solution specification form** can contain information such as formulation instructions, final solution appearance, pH, viscosity and specific gravity.

The format of a specification should be simple. Most specifications will be one page long. A second page becomes necessary only when testing requirements are extensive or when there are printouts or standard tracing attachments (such as HPLC profiles used for comparison in identity testing). Because testing methods are usually documented in standard operating procedures, when a specification document cites a testing method a simple reference to the SOP number is sufficient.

As a general rule, SOPs should tell how to perform the analysis, and the specification should describe an acceptable result. Exceptions arise, of course. If a method is short and will be used *only* for the release of this part numbered item, then it may be convenient to describe that method in the specification.

Every part numbered item received in the facility will be inspected by material handling and QC personnel to ensure that the item has not been damaged in transport; that the item ordered is, in fact, the item received; that it has been purchased from an approved vendor and that the part number assigned accurately describes the item.

Although every part numbered item must be visually inspected before it is released from the quarantine area, not every part numbered item requires testing before release. Determining which items will require testing, how they will be tested and what will be the test-acceptance criteria is one of the most difficult and demanding responsibilities of the QC manager.

Items That Require Testing

All critical chemicals and components must be tested before they are released; this requirement is mandated in the Code of Federal Regulations, Title 21 (21 CFR), Subpart I 211.160 and Subpart H 606.140a. Critical items are those chemicals or components that are a direct part of the final product or come in direct contact with the product during processing.

In addition to the testing required for critical items, any item subject to deterioration should be a candidate for testing. Also, any item purchased from a new or questionable vendor should be tested until that vendor can be properly qualified (discussed below).

Our Laboratories, Inc. Page 1 of 1

Specifications for Chemicals

Part Number: 6328

Item: Sodium Phosphate, mono, ACS   **Specification Edition #:** 01

Supplier	Approved Vendor	Catalog #	Quantity/Unit
Sigma	Sigma Chemical	S 9638	500 grams
Baxter S/P	Mallinckrodt	7892-500	500 grams
		789202.5	2.5 kg

Appearance: White Crystals

Chemical Formula: $NaH_2PO_4 \quad H_2O$

Formula Weight: 137.99

Storage Conditions: 20-25 °C

Precautions in Handling: none

Expiration Date: 5 years

Sample Size: 1 gram

File Sample: none required

Testing Requirements:

Test for sodium (SOP Q111): positive

Test for phosphate (SOP Q111): positive

Approval Signatures:

_____   _____
Quality Control Manager   Date

_____   _____
Verification   Date

Form QC142
Edition 02

(Figure 2.1)

The Extent of Required Testing

Critical chemicals and components require testing according to 21 CFR, Subpart E 211.84d and Subpart H 606.140b. Critical items should be tested for identity, purity, strength, stability and quality. If the vendor's certificate of analysis is acceptable (after the vendor has been qualified), then at least one identity test must be performed. Generally, a commitment to test a product's critical chemicals and components is made in official regulatory documents. When working under an investigational new drug (IND) application, new drug application (NDA) or product license application (PLA) ensure that these commitments are met in the specification documents.

Processing intermediates produced in-house—such as crude extract materials, purification intermediates and crude cell harvests—can also be tested as critical materials. These can be assigned critical chemical part numbers, and specifications can be developed to ensure that each production intermediate is tested according to its specification before it is released for further processing.

Other items also may require testing. Determine how much testing by considering the following questions:

- Does the item come in direct contact with the product during processing?
- Would failure of this item affect the quality or safety of the final product?
- What is the likelihood of failure?
- What is the potential seriousness of failure?
- How easy is it to test or ensure the quality of the item?
- How reliable is the vendor?
- Will a certificate of analysis accompany every shipment?
- Can the current QC staff support the testing commitments?
- Is it better to evaluate the quality of the item just before using it (through process controls)?

It is the responsibility of QC to worry about all of these considerations and to make testing decisions that can best safeguard the products produced at the facility. If one is inclined to worry too much and commit to too many testing regimes, then there is never enough time or money to support the operation. Decisions must be made. Be realistic in the decision-making process, but never stop asking "what if?"

When No Testing is Recommended

Many laboratory chemicals used in preparing assay reagents may require no testing. These items receive part numbers to ensure that chemicals of reliably equivalent identity and quality are used in analytical testing. Identity can be performed initially to qualify a vendor, or it can be performed periodically or not at all. These decisions can be made only by the QC manager who understands the items' ultimate use and impact on product quality and operations.

For example, a disinfectant solution concentrate used in clean room areas and purchased from a vendor should be qualified by the microbiologist before it is approved for use in the facility (disinfectant effectiveness testing), and its effectiveness should be routinely monitored (process control). Except, perhaps, for a bioburden assessment, little useful testing of the concentrate can be done upon its receipt. Nevertheless, in all cases part numbered items are inspected upon receipt to ensure that they meet the specification requirements. In this way identity is ensured, if not tested.

Testing Method Resources

The primary resources for information on testing critical raw materials, processing intermediates and final products are regulatory documents that have been submitted to the Food and Drug Administration (INDs, NDAs, PLAs). In addition, 21 CFR 210, 211, 606, 610, 680 cite mandatory testing requirements for drug and biologic products.

Vendors are also excellent sources of information on product characterization and testing methods. Many catalogs contain sections on product specifications and give references. In addition, if the vendor supplies a certificate of analysis, ask for a copy of a testing method that will allow periodic confirmation of results. Some vendors are very helpful.

For most chemical entities, section 191 of the *U.S. Pharmacopeia* describes testing methodologies. These wet chemistry tests are inexpensive and easy to perform. Their use, however, can only add to the supply of chemicals and reagents in the QC laboratory, which in turn means more specification writing. Although some laboratories can afford sophisticated, analytical, molecular identification systems that save technician time, validation and training requirements should also figure into the purchasing decisions.

Other resources for specification writing include *The Merck Index* (Merck & Co., Inc., Rahway, New Jersey), *Standard Methods for the Examination of Water and Wastewater* (American Public Health Association, Washington, D.C.), and *Official Methods of Analysis of the Association of Official Analytical Chemists* (AOAC, Arlington, Virginia).

It may be impractical to test some items upon receipt. If a critical item is not tested upon receipt, however, then the QC or QA manager should evaluate the potential need for process controls (testing) on that item before it is used in a critical situation. For example, if media are purchased in powdered form, it might be impractical to test them for growth promotion or cytotoxicity upon receipt. Instead, it may be established that every batch of fully formulated media will be tested for growth promotion, osmolarity, sterility and endotoxin.

Setting Acceptance Criteria

A manufacturing facility can either accept the limits set by a vendor or establish more stringent limits. When you have no idea where to start, pick a reasonable value, one that will not negatively affect the quality of the process that the item supports. Then follow the item's testing history to determine whether or not the limit is reasonable.

Writing, Approving Specifications

Generally a specification is written when an item is ordered for the first time. To help the QA manager write specifications, design a part number request form—to be completed whenever someone needs a new part number. It is helpful for the QA manager to consult with the individual who requested the item in order to fully understand how it will be used and what attributes of the item could adversely affect product or process.

When an original specification is completed, both the requester or user and the QA manager can sign it. At this time a specification file can be assembled.

Assigning Expiration Dates

Every specification should indicate an expiration date for the item. The entry should indicate the length of time, from receipt, that the item can be used and still maintain its original quality. A company policy should guide the expiration dating of items, by category. Exceptions will certainly occur; for example, when a vendor has assigned an expiration date to an item, that date should be honored unless the in-house date is sooner.

For routine dry chemicals, such as sodium phosphate and sodium chloride, a five-year expiration should be adequate; for solutions, three to six months. Pick an expiration date that is reasonable—ideally a time before the material deteriorates and after it is used up. Consult with the company's financial advisers when stock chemicals are expensive or if it seems likely that much of the material will remain when the expiration date is reached.

For sterile products, the expiration date can depend on storage conditions. These expiration dates should therefore be validated. When critical items are stored sterile, for example, it is advisable to confirm that they can be stored for the indicated time without compromising their sterility. Such process control programs can only strengthen specification testing decisions.

Sample Size, File Sample Size

The size of the sample required to evaluate a part numbered item is determined by the amount required to perform the testing, the amount or number received, and the relative expense of the materials. If a 500-gram bottle of sodium chloride requires testing, for example, then it is good practice to give the technician twice the amount required to perform the analysis.

When a large number of identical units must be sampled, however, a sampling plan must be established and followed (21 CFR 210.3). The MIL STD (Military Standard) 105E, "Sampling Procedures and Tables for Inspection by Attributes," is one resource for sampling plans. It is available from the Standardization Document Order

Desk, 700 Robbins Avenue, Building 4, Section D, Philadelphia, Pennsylvania 19111-5094.

File sample or reserve sample size is specified in 21 CFR, Subpart I, 211.170, which indicates that the company must save twice the number of units needed to complete full specification testing of the item. It also indicates that, when determining the number of reserve samples of final product, it is not necessary to include the number of samples needed for sterility and pyrogen testing. This requirement is product dependent, however, so it is necessary to consult product IND, NDA or PLA documents.

Vendor Qualification, Approval

Vendors of critical chemicals or components must be qualified. This means that QC or QA must ensure through any means possible that the vendor's manufacturing process can reliably and consistently produce an item of known quality. In addition, vendors must assure purchasers that they will report any major changes in the manufacture of that item.

Vendors can be approved in several ways. Thorough testing of an item is one way to ensure its quality. Sometimes, however, rigorous testing is too difficult or too expensive. Another way to ensure quality is to request a certificate of analysis from the vendor with each lot and to confirm periodically the vendor's results. Finally, a vendor audit is the best way to ensure quality, and it is likely to result in the kind of direct communication necessary to an open working relationship.

Vendors can also be analytical testing laboratories that help perform specification testing. It is important to qualify their operations through on-site audits and by occasionally sending them spiked samples.

Specification as a Commitment

An approved specification becomes a commitment to test the specified item as indicated. If a company is unable to fulfill the commitment of the specification, then it should not be written. During the start-up phase of an operation, there may be times when items cannot be as thoroughly tested as they will be at a future date. When that is the case, indicate it on the specification, and change the specification when it becomes possible to upgrade the testing facilities.

Specification Files, Specification Books

A specification file is literally a paper file (such as a green hanging folder) that contains several subset files. The subset files include the current (original) approved specification and all former specifications for that item; documentation of all raw data testing that has been performed on that item over the years, filed by receiving code; purchase order and receiving documentation on that item (optional), filed by receiving code or lot number; product information from vendors; and any other information specific to the item.

These files should be kept in numerical order in a secure but not restricted location. The specification files are a primary source of information for QC, and routine access must be ensured; they are usually maintained by QC or QA personnel.

A specification book is a three-ring binder that contains all current specifications in numerical order; it must kept be in the QC laboratory. Additional copies may be appropriate for receiving or production areas. Because those copies contain the actual reference copy of the specifications that technicians use from day to day to release items, they must be controlled and updated immediately when a specification changes or a new one is added.

Changing Specifications and Edition Numbers

Specifications for items will inevitably change. If it is a critical item, the change may require FDA approval before it can be instituted. If a change in a critical item specification adds more testing requirements to the current specification or tightens an existing limit, then change can usually proceed without immediate FDA notification.

Change to noncritical items is not regulated. However, change must be controlled. If a specification change is significant, resulting in a different item or grade of chemical, then it is appropriate to assign a new part number to the new item, write a new specification and retire the old part number. If the change is minor, then it is enough to change the specification and issue a new one to all specification notebooks—retrieving and destroying the previous version. In the specification file, put a line or an X through the old specification, sign and date the cross-out and place the new specification in the front of the folder.

The new specification is now a revision of a former specification, and the revision must be documented. Edition numbers, sometimes called revision levels, accomplish this. They are simply numbers (01, 02, 03, 04) or letters (A, B, C, D) that must appear on every specification. Develop a document history form to record edition numbers and the changes in that document over time.

Each document requires two numbers to be identified completely: a document number (SOP number, form number or specification number) and an edition number. The document number tells what it is, and the edition number tells which one it is.

Operational Suggestions

As mentioned above, several forms are required to facilitate the use of this system. Forms should be designed for assigning part numbers (which are used to draft the specification). Specification forms should be designed for each individual category of parts (chemicals, components, cell lines, printed materials, solutions, final product). A specification history form should detail the changes to that specification over time and associated edition numbers.

A specification notebook should be available in the QC laboratory, and specification files should contain the entire history of material receipt and testing. SOPs should be written on specification writing, approval, use and change. There should also be a written commitment to audit the specifications annually.

An Overwhelming Task

During start-up, the task of writing specifications and performing all the release work required as items arrive seems nearly impossible. The only way to manage this situation effectively is to make good testing decisions. Decide what absolutely requires testing and what can wait until more equipment and personnel are available. Also during start-up, alliances with outside testing laboratories can facilitate the release testing that cannot, as yet, be completed in-house. Until you have the time and the support to worry about everything, worry first about the things that matter most.

3

Lot Numbers

THE ACCOUNTABILITY AND TRACEABILITY fundamental to Good Manufacturing Practices cannot be achieved with part numbers alone. To identify an item completely, both the part number and a lot number are required. The part number tells *what* item it is, and the lot number tells *which* item it is. Unless both numbers appear on an item, its identity is questionable.

Lot numbers are assigned when an item is received into the facility (receiving codes) or when an item is produced within the facility (in-house lot numbers). The assignment of these numbers must be documented and controlled.

Receiving Code Assignment

When a part numbered item is received into the facility, the receiving clerk documents its arrival. The clerk matches the item to its purchase order, locates the part number on the order and records receipt of the item in a receiving log book.

The receiving log book contains column entries for the part number, a description of the item, amount received and its configuration, supplier, manufacturer, manufacturer's lot number, purchase order number, initials of the individual logging in the item and comments. In addition, items are assigned receiving codes as they are logged into the book.[1]

Receiving codes—simple numbers or alphanumerics—are prelisted in the receiving log book. Receiving codes of AA001, AA002, AA003 . . . AA999, AB001, AB002, etc., are adequate for most operations. When a bottle of sodium phosphate, monobasic,

ACS grade, arrives, for example, and the purchase order indicates that the part number (PN) is 6328, the receiving clerk enters information from this shipment on the line in the log containing the next available receiving code (for example, AA303). The bottle is then labeled with both the part number and receiving code.

A separate receiving code should be assigned to each separate manufacturer's lot and for each separate shipping event. For example, twenty bottles of sodium phosphate (PN 6328) arrive, fifteen with a manufacturer's lot number of 4R89T and five with a manufacturer's lot number of 5Z89N. The fifteen bottles should be assigned a different receiving code than the five bottles. If another shipment of five bottles of sodium phosphate (PN 6328) arrives the next day with a manufacturer's lot number of 5Z89N, these five bottles should be assigned a third receiving code.[2]

In order to assure the control and documentation of incoming materials, all materials received into the facility must be routed through a central receiving department. This includes packages delivered to the receptionist by the postal service or overnight courier and packages that come in after hours.

Labeling

Each item received is labeled with its part number and receiving code and then placed into a locked quarantine area. In addition to part number and receiving code (RC) entries, the quarantine label contains an optional description of the item, an indication of how many units were received under this RC (1 of 4, for example) and a "sampled by" and "date" entry. The word "quarantine" should be at the top of the label, which is customarily an orange color. When the item is labeled and moved into the quarantine area, QC is notified of its receipt. The item can then be inspected and, if necessary, sampled and tested.[3,4] (Figure 3.1)

When the item is released, QC prepares a released label and places it next to the quarantine label. The release label, which is customarily green, may also contain the part number and receiving code of the item. It should also show the expiration date and storage conditions for the item. Material handling personnel then move the item to a release area. It may also prove advantageous to record the release in the receiving log by adding a "release date" column. (Figure 3.2)

If an item is rejected as a result of QC inspection and/or analysis, it receives a reject label—customarily red in color—and the item is moved into a locked reject area to await disposal or return. (Figure 3.3)

Solution Preparation Lot Numbers

Solutions are often prepared daily in several areas of a manufacturing facility. All solution preparation events must be controlled and documented; there are several ways to do this. One method calls for solution log books in the production area and in the QC laboratory to record, in chronological order, every solution prepared; in-house lot numbers may be assigned from this book. Log book entry columns should contain a

QUARANTINED

Part # 6328   RC # AA303

Item # Sodium Phosphate Monobasic, ACS

Configuration: 1/4   units

Sampled by:_____ Date: 9/12/91

RELEASED

Part # 6328   RC # AA303

Storage Conditions: 20-25 °C

Expiration Date: 8/98

REJECTED

Part #_____   Lot/RC #_____

Date:_____   Technician:_____

(Figures 3.1, 3.2, 3.3)

description of the solution and/or its part number, the date of preparation, the volume prepared, pH (if appropriate), concentration, and solvent used (if other than water), along with the name of the preparer and any comments.

Alternatively, individual solution preparation forms can be designed to document each event, with a solution lot number log book to control and document the assignment of lot numbers. If this system is used, ensure that all information listed for the solution log book appears on the form. (Figure 3.4)

Many facilities simply use the date of preparation as the in-house lot number. However, this becomes inadequate if it is possible for several departments to prepare the same solution on the same day. If this is a concern, design a solution lot numbering system that identifies where the solution was prepared (QC, production, research), then gives the date and/or some chronological numeric listing. Ensure that the format of this in-house lot number is different from that of the receiving code or production batch lot numbers. For example, a solution prepared in the QC laboratory on 8/14/91 could be labeled with lot number Q910814.

Whatever the system, each solution must be labeled with its complete name and/or

DOCUMENTATION BASICS

Our Laboratories, Inc. Page 1 of 1

SOLUTION PREPARATION
(SOP 572)

Solution: 0.1 N Sodium Hydroxide   **Part #** 8012

Final Solution Lot #: 910712

Date of preparation: 7/12/91   **Prepared by:** _____

Expiration date: 10/12/91   **Storage conditions:** 20–25 °C

SOLUTION PREPARATION

Prepared according to specification #: 8012 ; **edition #** 02

#	Units	Chemical	Part #	Lot #/RC #
4.0	grams	Sodium Hydroxide, ACS	6053	AB123

Solvent used: ( ) purified water   (x) WFI
   ( ) Other _____ Part # _____ Lot/RC # _____
   Final volume = 1000 mL

pH adjustment required?   ( ) YES   (x) No
   _____ mL of Part # _____ / Lot/RC # _____
Final pH _____

SOLUTION QC TESTING   (x) none required   (see below)

_____
_____
_____
_____
_____
_____
_____
_____

          Technician _____ Date _____

Form 4/153
Edition 02

(Figure 3.4)

part number, date of preparation and/or lot number, storage conditions, expiration date and the preparer's name. Labels can be printed to facilitate proper and consistent solution labeling throughout the facility.

Written instructions should specify how to prepare solutions for general use. A convenient place to write these instructions is in a solution part number specification document, readily available for reference in the QC laboratory and production preparation area. This document contains information on testing the solution, storage conditions and expiration dating.

When a solution is prepared for a single use only, its preparation can be documented directly and solely on the assay form or in the batch record. For example, if a buffer solution must be prepared fresh for a potency assay, the first page of the assay form can contain a section on solution preparation. This section would document the same information that is contained in the solution log book.

Cell Line Lot Numbers

Cell lines can be received from outside sources or developed in-house. In either case, it can be useful to route a cell line through the receiving system when the line is finally available for use in GMP production. This ensures that the cell line is labeled with a part number and a receiving code and evaluated according to a specification before it enters the production area.

For example, a customer sends a cell line (two vials) to the facility in March for use in producing a product. This cell line part number is 3015. When it is received, a receiving code of AB665 is assigned; the vials are labeled (discussed below) and transferred to liquid nitrogen storage in a locked quarantine area pending QC release.

When the cells have been released, it is common practice in this example facility to create a working cell bank (WCB) from the master cell bank (MCB). This is done by removing a vial and expanding the cells to create a series of identical vials for the WCB. QC reevaluates these WCB vials before releasing them for use in upcoming production. When a production run is scheduled, one of the vials from the WCB is used for inoculum preparation.

In September of the same year, the customer sends two more vials of identical cells to this example facility. These are also received as PN 3015, but are assigned a different receiving code—AB901. Another WCB is created from this MCB vial.

Clearly, an MCB cell line is different from a WCB cell line, and this difference must be easily recognized from the label. It should be designated in the part number because the MCB and the WCB are distinct entities that are tested to meet established specifications and then stored until required for further processing.

To facilitate this, an "M" or a "W" can be added as a suffix to cell line part numbers. Separate specifications can describe the testing requirements for each. In the example above, the March shipment would be labeled "PN 3015/M, RC AB665/1" and "PN 3015/M, RC AB665/2." Subsequent WCB vials created from "PN 3015/M,

RC AB665/2" would be labeled "PN 3015/W, RC AB665/1." Similarly, a WCB vial prepared from the September shipment might have a label such as "PN 3015/W, RC AB901/1." If one needs to know exactly which MCB vial was used to create any individual WCB vial, this information should appear on the first page of the form that documents the WCB expansion event.

Cell identity issues. Oftentimes cell lines are used in a process that continues until, for example, there are enough cells to inoculate a bioreactor. In such cases several major scale-up events and, therefore, cell identity issues must be controlled and documented. Cells are transferred from T-flasks to roller bottles and fermenter tanks as their numbers increase. Because no point in the scale-up process is a stopping point, any designation that differentiates one cell from another—such as passage levels (P1, P2, and so forth)—should be associated with the lot number or receiving code number, not the part number. If a vial number is recorded on a P5 passage form, for example, the labels should read "PN 3015/W, RC AB901/1/P5" or "PN 3015/W, RC AB901/P5." Any testing performed on the cells during this scale-up event can appear on forms in the batch records that accompany the process. A part number specification document is not appropriate for this.

When a cell line is used to produce a product, the cell is simply raw material. As soon as a cell line is inoculated into a bioreactor or fermenter, its purpose is to produce product. The process is therefore identified by the product it produces, rather than the cell line used to produce it (discussed below). The first page of a batch record for the inoculation of a bioreactor, however, must specify the cell line used (for example, "PN 3015/W1, RC AB901/P8"). Batches and batch records will be presented in a forthcoming chapter.

Production Lot Numbers

Because every part numbered product or process intermediate is manufactured in a given facility many times throughout a year, lot numbers serve to distinguish between identical processing events. Therefore, a lot number is assigned to each separate production event for each distinct processing intermediate or final product manufactured in a GMP facility. The lot number provides information, within its format, about time and location parameters of manufacturing.

A production lot number, also called a batch number, is defined in 21 CFR 210.3: "Lot number, control number or batch number means any distinctive combination of letters, numbers or symbols, or any combination of them, from which the complete history of the manufacture, processing, packing, holding and distribution of a batch or lot of drug product or other material can be determined."

A batch is also defined: "Batch means a specific quantity of drug or other material that is intended to have uniform character and quality, within specified limits, and is produced according to a single manufacturing order during the same cycle of manufacture."

A production lot number is the most public number generated in a GMP facility;

it appears on final product that leaves the facility and circulates in the consumer market. If problems arise with a product, this is the number cited on any complaint. The production lot number, as described in the CFR, must enable the producer to trace through records and produce all documentation that supports the production, testing, stability, inspection and packaging of that product. The batch record facilitates this traceability, and the lot number uniquely identifies that batch record.

A production lot number for a producer with more than one facility approved to produce the product could be formatted as "C2A911," for example. Here the first letter describes the location of the facility that produced the batch; in this case, C = California, K = United Kingdom, P = Puerto Rico. The next character, a number, indicates which of the identical machines or lines validated for manufacture of this product was used to produce this lot. The following letter indicates the month in which production was initiated; A = January, B = February, C = March. The next two numbers indicate the year, and the final number indicates chronology; in this case, the first batch initiated on that machine or line in that facility in that month of that year.

Production lot numbers should be assigned by the person in charge of scheduling production events. A log of these numbers and their assignment must be maintained.

Harvest, Harvest Pool, Purification Numbers

Given a production batch number of PN 4015, lot number B2A911, how does one thoroughly label harvest fluids, pooled harvest fluids and subsequent purification intermediates? Although this is a decision that, again, is unique to every operation, each processing intermediate must be identified primarily as a component of the 4015-B2A911 process and must, at all times, carry that designation.

An alphanumeric system can further identify and track the process intermediates of a batch. The following scenario is offered as an example. Over the course of a 30-day production event, 45 individual harvesting events are performed. Each harvest is filtered immediately to remove cells, and harvest filtrates are stored at 2-12°C for not more than six days or until 2-3 liters of fluid are collected. Pooled harvest filtrates are processed through an initial purification step and held again at 2-12°C for up to 30 days until enough material is collected to perform the final purification steps.

Labeling the intermediates. As already mentioned, the part number and lot number of the production event (4015-B2A911) must appear on every container. In addition, crude harvest materials can be labeled with a processing identifier such as CH1, CH2, or CH3, and harvest filtrates can be labeled with a processing identifier such as HF1, HF2, or HF3. (Figure 3.5)

The beginning of purification is usually formatted into a new section of the batch record. A bill of materials appears on the first page of the batch record; it requests the identity of all harvest filtrates pooled in order to proceed with this purification step. The material resulting from this initial purification (IP) can be labeled 4015-B2A911/IP1. The next time this initial purification step is performed for this batch, it will be

```
┌─────────────────────────────────────────────┐
│           PROCESS INTERMEDIATE              │
├─────────────────────────────────────────────┤
│ Product #  4015      Batch #  B2A911        │
│ Process Intermediate Identifier: Harvest filtrate-1 │
│ Configuration: 1/3 units  Approx. vol/wt: 1 liter │
│ Storage Conditions: 280 °C                  │
│ Expiration Date: 9/16/92                    │
│ Technician: _____ Date Collected: 9/2/92 │
└─────────────────────────────────────────────┘
```

(Figure 3.5)

labeled 4015-B2A911/IP2, and so on. Similarly, final purification product can be labeled with a processing identifier such as FP1 or FP2.

These processing identifiers are in-house numbers that facilitate control and traceability of materials. When the purification is complete, and QC testing indicates that product 4015 meets the acceptance criteria of its specification, the final product can be labeled simply "4015-B2A911."

Processing intermediates are identified because they must be controlled. Typical process controls set limits on how long an intermediate can be stored before proceeding to the next step and on the maximum number of units that can be pooled to prepare for the next step.

Routine Use of Part Numbers, Lot Numbers

When writing SOPs and master batch records, cite the part numbers of all items described in the narrative. When designing a form or a fill-in-the-blank section of a production batch record, cite the part number of the item and provide a blank space for the technician to fill in the receiving code or in-house lot number when the operation is performed.

Operational suggestions. SOPs should be written to support policies on lot numbering, receiving, labeling and handling of rejected materials. Log books should be available and controlled for the assignment of all lot numbers. Forms should be designed for the sampling, inspection and testing of part numbered items. The format and content of SOPs and forms will be presented in the next chapter.

When labels are first designed, it is often too costly to have them printed. A convenient alternative is to buy blank colored labels and have a rubber stamp prepared that contains the information required on the label. Simply stamp a stock of blank labels and proceed. When it is determined that the label seems to function properly, a permanent supply can be printed.

Summary

Clearly, there are many ways to create a lot numbering or part numbering system. However, the system must work to support the GMP requirements of accountability and traceability of materials. Beyond that, a good system is convenient, simple, informative and easy to understand.

A part number should only identify an item, telling "what it is." A lot number should only differentiate identical items from one another by parameters of time or location, telling "which one." To identify any item completely requires at least two numbers—the part number and a lot number (or receiving code). Processes are defined by the products they produce; a stopping point in a process where product is tested and stored for future use is the end of that process, and the product should be labeled with a part number and meet its specification for acceptance. If within a day or two the process continues beyond the creation of this product intermediate, then label the intermediate with the part number and lot number of the final product and a processing identifier.

As you reflect on your current systems and their inevitable awkward points, consider the basic principles presented in this book. These basic guidelines can help clarify why your systems are awkward and show how you can revise them—if you are brave.

4

Standard Operating Procedures

STANDARD OPERATING PROCEDURES (SOPs) are documents that describe how to perform various routine operations in a GMP manufacturing facility. They contain step-by-step instructions that technicians in QC, production, maintenance and material handling consult daily in order to complete their tasks reliably and consistently. SOPs are, in essence, written commitments to the Food and Drug Administration that describe the performance of routine tasks. An FDA inspection evaluates how well these written commitments are fulfilled.

Data collection forms are documents a technician completes while performing routine tasks. The forms often recite the step-by-step instructions of the SOP while providing fill-in-the-blank spaces for the collection of raw data entries.

SOP Format

There are many ways to format an SOP. When developing a format, it is important to consider a number of informational categories: title, purpose, scope, responsibility, references/applicable documents, safety considerations, procedural principles, preliminary operations, procedures, calculations and documentation requirements. (Figure 4.1)

The **title** in an SOP format should be brief and direct, describing each procedure in a way that identifies its purpose. The title should also include any key words useful for locating the procedure in a list of SOPs. Titles can be used to group SOPs according to function. For example, many SOPs are required to describe the operation of equipment in a facility. If these SOPs are titled "Operation of . . . ," they could be easily located on an alphabetical listing.

Our Laboratories, Inc.

Page 1 of 1

Final Product Sterility Testing
(SOP 97)

Product _____ Batch/Lot # _____

Test Date _____ ; # Units tested = _____

Vials reconstituted with _____ mL of _____

Thioglycollate media PN 6711 RC # _____ ; or Media # _____

Soybean–Casein media PN 6712 RC # _____ ; or Media # _____

Peptic digest rinsing fluid media # _____

Sterility test apparatus filter units PN 7044 RC # _____

Day	Thioglycollate media Incubation 33–37 °C*		Soybean–Casein Incubation 20–25 °C*		Technician
	Sample result	Control result	Sample result	Control result	
1					
2					
3					
4					
5					
6					
7					

*Record growth as   and no growth as

Settling Plate PN 6998 RC # _____ Incubation results = _____ CFU

RESULT: ( ) No growth observed in sample bottles OR negative controls after 7 days; SAMPLE PASSES STERILITY TEST
            ( ) Growth observed in sample bottles; DOES NOT PASS
            ( ) Growth observed in negative controls; TEST INVALID

Technician _____ Date _____ Verification _____

Form # 4/023
Edition 02

(Figure 4.1)

Purpose in an SOP format often restates a well-written SOP title, but it can also be used to expand upon or qualify the purpose of the procedure.

Scope describes what the SOP does and does not apply to. For example, if an SOP describes the calibration of a spectrophotometer, the procedure might include only spectrophotometers in the QC laboratory, only spectrophotometers used in GMP operations, only spectrophotometers used for potency assays, or only double-beam spectrophotometers. To declare the scope of an SOP, therefore, consider to what and to whom the procedure applies, and when it is to be applied.

Responsibility in an SOP format simply declares who is responsible for performing the operations cited. It might cite a department or mandate specific training requirements for individuals within a department.

The **references and applicable documents** sections of an SOP are optional. They may cite the origin of the procedure, such as the journal reference for an analytical assay. These sections can also be used to reference allied SOPs, protocols, batch records, or further information sources, such as vendor manuals and instruction booklets.

Safety considerations should appear in all appropriate SOPs. These include physical safety issues (hard hats and goggles, for example), biological contamination issues (such as masks, gloves and biological safety cabinets) and chemical hazards. Cleanup of any spills that could occur during a procedure should also be described here.

Procedural principles/introduction, although optional, is most appropriate for analytical assay SOPs. It is provided to help technicians and reviewers understand the fundamental principles of the assay. This section also can be used to explain why the procedure is required in the context of other facility operations.

Preliminary operations is another optional section of an SOP. It can include any operations that should be completed before the actual procedure is initiated. In an assay SOP, for example, there may be a need to prepare solutions or calibrate equipment before the assay begins. In addition, some facilities require a material checklist section to ensure that all materials required to complete the work are available before the work begins.

Procedures should be simple, direct, step-by-step narratives that explain how to perform the tasks in a manner that supports process control. Any sampling or testing that might occur to support the task should also be cited. Examples are discussed later in this chapter. Include diagrams and drawings in this section when they facilitate understanding of the instructions.

Calculations of a final result must also appear in appropriate SOPs. These instructions should explain the calculations completely (no unexplained "magic numbers" or formulas). Examples are helpful. Although a procedure should explain how to calculate a final result, generally it should not dictate acceptance criteria for that result. Specifications or protocols are the documents that establish acceptance criteria; protocols are discussed in upcoming chapters.

Documentation requirements for an SOP should cite any log books, batch records or forms to be completed during the procedure. This section should address distribution of results, if appropriate. Procedural deviation documentation requirements should also be outlined or referenced in this section.

Each page of an SOP document should contain its abbreviated title, SOP number, edition number and pagination (for example, page 1 of 12). The company name and some declaration of confidentiality can also appear on each page. As long as each page contains the SOP number and its edition number, only one page—usually the first—needs to contain the document approval signatures and the date of approval.

Who Writes SOPs, When?

SOPs support routine GMP-regulated operations. They are written when it is clear how the task will be performed (procedure), who will perform it (responsibility), why it will be performed (purpose) and what, if any, limits of use apply (scope).

SOPs should be written by or with the individuals who perform the operations. Although particular individuals in each department may write the bulk of the procedures, it is important that everyone be trained to draft and revise SOPs.

SOP Language and Detail

Do not write SOPs for the FDA. Write them for the technicians who will use them on a daily basis. Use clear and direct language. Use active verbs for procedural directives: "add this," "pour that," "observe color." When citing the use of an item in a procedure, name the item and its part number. Company slang terms for equipment, departments, or procedures can be used, as long as they are explained somewhere in the text to ensure clarity for an outside reviewer such as the FDA. It is good policy for draft SOPs to be reviewed by the technicians who will use them.

SOPs must be specific enough to be clear and accurate, yet flexible enough to be useful. Too much unnecessary detail can render a procedure useless before it is even approved. A directive can be specific, yet truly uninformative. Here is an example: "Place the test tubes in the water bath in Room 28 for 30 minutes. Spin the tubes in the IEC centrifuge at setting #5 for 20 minutes. Pour off the supernatant and dialyze it for 24 hours in cold WFI." Although technicians may be able to perform those tasks as directed, any change—even a minor change—will require a change in the SOP. Furthermore, important details that support process quality control are missing from these instructions.

A directive such as "Place the test tubes in the water bath in Room 28 for 30 minutes" should be changed to "Heat the tubes in a water bath to 35-39°C for 30 minutes (+/- 5 minutes)." The action directed by this statement is to heat the tubes for an established amount of time at an established temperature. This is all that needs to be directed. It must, however, be directed specifically.

"Spin the tubes in the IEC centrifuge at setting #5 for 20 minutes" is also too detailed, yet insufficiently specific to ensure good process control. The intent of this step is to subject the tubes to an established amount of centrifugal force for an established period of time. Although that is accomplished at setting #5 in this IEC unit, there is no flexibility for the use of another, equally adequate centrifuge. Instead, state, "Spin the tubes at 8,000-10,000 xg for 20-25 minutes at 2-8°C (example: #5 setting on IEC centrifuge A267)." The routine use of the IEC centrifuge can be acknowledged in the SOP when it is presented as an example that supports the true xg criteria. In this way, one indicates process controls on temperature, time, and gravitational force without committing to too much unnecessary detail.

SOPs for dialysis events should indicate the part number of the dialysis bags (or membranes) approved for use, as well as a way to know when the dialysis is complete, such as conductivity of final dialysis water. For example, "Pour off the supernatant and dialyze it for 24 hours in cold WFI" is specific, but not in a way that supports process control. How would one know, in this example, when dialysis is complete?

Although temperature parameters must be established for dialysis events to ensure product stability, time parameters are related to the rate of dialysis, and therefore to water volume and exposure time. Consequently, it is more appropriate to indicate the dialysis end point, then cite an example of water volume and time requirements. For example: "Dialyze not more than one liter of supernatant fluids in bags (PN 2005) until conductivity of dialysis water is equal to or less than WFI (example: 20 liters of dialysis water changed every six hours should require not more than four changes)."

SOP Number Assignment

SOP numbers should be assigned and controlled by one individual in QA. SOP identity numbers can be simple numbers, such as 572, and may be categorized. Number assignments should be kept in a log book, indicating the SOP number, the date of issue, the requester, a preliminary title and the proposed content. The SOP author may put this information on a form and submit it to the QA manager. After reviewing the form, the QA manager might suggest revising an existing SOP to contain the new information instead of issuing a new SOP number. The QA manager must also evaluate what impact, if any, the new procedure will have upon existing procedures in the facility and must initiate any changes in the system required.

SOP Review and Approval

There should be at least two signatures on an SOP, and preferably three. The person who wrote the SOP should sign it first. Then a person who can knowledgeably review and approve the procedure (often someone else in the originating department) should sign. Finally, QC or QA must review and sign all SOPs. Many companies require additional signature blocks on a procedure. It is advisable to minimize the number of signatures required and to maximize the ability of signatories to effectively review the SOP.

SOP Distribution and Control

The distribution of approved SOPs must be controlled. If each department receives a master set of procedures, the periodic distribution of new procedures to these departments must be documented. Whenever a revised procedure is issued, it must also be documented that the old editions of the procedure were retrieved and destroyed. There are many ways to document these events with either log books or forms. Whatever the system, one should be able to track the history of creation, change, distribution and use of each document.

SOPs must be in the areas of the facility where the work they describe is performed. A complete set of SOPs may be on file in the QC manager's office, for example, but copies of these procedures must also be in the laboratory or in the equipment room where they can be used daily. If there is an SOP on the operation of the autoclave, a copy of that procedure must be kept near or on the autoclave.

SOP changes. The apparent universal reluctance to write SOPs may stem from a perception that, once written, they cannot be changed. Change, however, is inevitable and SOPs must be changed to accurately reflect the work as it occurs. This change must be facilitated so that it can occur quickly and easily. If it is too difficult or too time-consuming to change an SOP, technicians and managers alike will find a way around the system, placing both process and facility out of compliance.

Only when a change in an SOP document significantly alters a critical processing event will approval of corporate regulatory authorities be required before the change can be instituted. This generally applies to procedures and techniques that support current applications to the FDA.

All SOPs are assigned edition numbers that reflect the revision level of the document. The first issue of an SOP could, for example, have edition number 01. When a change occurs—even a minor change on just one page of the document—the entire SOP is reissued as edition 02. As the document is revised over time, a history of change must be maintained indicating what changes occurred, why and when. Every edition of an SOP should be maintained in documentation.

SOP Topics

The initiation, approval, distribution, use and change control of SOPs and data collection forms must be described in a written procedure. Additional required topics for SOPs appear in the Code of Federal Regulations, Title 21, cGMPs for finished pharmaceuticals (Parts 210, 211), and cGMPs for blood and blood components (Part 606). There are at least 25 separate citations to written procedures. Review this list and consult other listings to draft a list of SOP titles.[4,5]

Data Collection Forms

As stated above, data collection forms are documents completed by a technician while performing routine tasks. The forms often include the step-by-step instructions of the SOP while providing fill-in-the-blank spaces for the collection of raw data entries.

In traditional research laboratories, laboratory notebooks are used to record data. The entries are informal—seldom are reagents identified completely, rarely is equipment identified, and signatures are missing. This is unacceptable for GMP manufacturing. Forms are created, therefore, to ensure that all information that must be documented while performing a procedure is, in fact, recorded.

Data collection form format. Forms should be designed to facilitate the work they are used to record. The entry blanks should appear in the order in which the work is routinely performed and should require the technician to do as little writing as possible. Short, fill-in-the-blank entries, checklists and lists of answers to circle are all appropriate formats. Fill-in-the-blank entries on a form are appropriate for receiving code entries or lot numbers, time, temperature, equipment identification numbers, room numbers where work is performed, calibration values, raw data values and calculations.

Each page of each form should contain the name of the company, the name of the form, an appropriate SOP reference, the form number, the form edition number, pagination and two signature and date blanks: one for the technician who performs the assay and one for a verification signature that indicates the form has been reviewed for accuracy, completeness and compliance.

Use and Control of Forms

After a technician completes a form, it is reviewed by a superior or someone else knowledgeable about the operation. The verification signature indicates only that the information recorded is accurate; it implies no judgment about the acceptability of the result.

Forms are controlled but not specifically approved. They can be written by the department responsible for the SOP or the task documented. Forms should be revised whenever a revision would clarify the data record or add to the information collected. There should be no prolonged approval process for form revision because significant information collected should be mandated in SOPs. A brief revision history of the forms, however, should be available.

Blank forms should be available for use at all sites throughout the facility where they might be required. Easy access to data forms helps to ensure that they are used.

In addition, a completed form may be copied for filing in more than one location. But the original forms must be kept in a known, secure location because FDA reviewers are likely to request originals for review.

5

Master Production Batch Records

BEFORE BEGINNING A DISCUSSION of batch records, it is important to consider the Food and Drug Administration's definition of batch, as cited in the Code of Federal Regulations, Title 21 (21 CFR 210.3): "Batch means a specific quantity of a drug or other material that is intended to have uniform character and quality, within specified limits, and is produced according to a single manufacturing order during the same cycle of manufacturing."

For a parenteral product, such as a 10-ml vial containing a liquid-filled lyophilized product, a batch of final product begins with formulation. The batch is then filter sterilized, filled aseptically, lyophilized, sealed, inspected and labeled. In all of these operations, all of the vials are exposed to the same conditions of processing at the same time.

For a cell culture–derived bulk product, a batch usually begins with inoculation and then proceeds through harvest and possibly through initial purification steps. Whenever the cycle of processing stops, the batch ends. If, for example, harvests are processed to remove cell debris and to concentrate product after harvest, and if the resulting process intermediate can be stored at -80°C for an extended period of time, this might be the end of the batch. If it takes several batches of cell harvest to provide enough starting materials to proceed through final purification, the final purification steps could be considered a separate production event with a separate batch record.

Once the batch is defined with a known beginning and end, and the separate processing events (final product formulation, filtration, filling, lyophilization, sealing, inspection, and labeling; or bulk product inoculation, monitoring and harvest and purification) are identified, a master production batch record (MPBR) can be designed.

The MPBR is a detailed, step-by-step description of the production process. It explains exactly how the product is produced, indicating specific types and quantities of components and raw materials, processing parameters, in-process quality controls, processing intermediate specifications, environmental controls, and so on. The MPBR described in this chapter is cited in 21 CFR 211.186 as the master production and control record.

To support the requirements of the production batch record (PBR), discussed later in this chapter, the MPBR should contain fill-in-the-blank spaces throughout the text to facilitate the documentation of events for each individual batch.

Production Batch Record

The production batch record (PBR) is an exact copy of an approved MPBR. It is issued for each batch of product produced in the facility. It contains instructions on how to produce the product, and provides space for written entries to document that each routine product batch was produced according to an established and approved MPBR. The PBR described in this chapter is cited in 21 CFR 211.188 as the batch production and control record.

MPBR Format

Each page of the MPBR should contain the company name, product name, part number, dosage and/or configuration, and, if appropriate, the site of manufacture. Each page should also contain a document number, an edition number, the theoretical yield of product associated with this MPBR and pagination. Spaces should be provided for the lot number of the batch and for the subsequent approval of the PBR; these two spaces will be completed when the PBR is issued.

The first page of the MPBR, or the first page of each of its sections, should contain original approval signatures, with dates, for the document. There should be at least two—preferably three—signatures from personnel in the production and QC departments.

It is convenient to divide the MPBR into sections that correspond to the major processing events of manufacturing. Appropriate MPBR sections for a lyophilized vial product would be component preparation (the cleaning and sterilization of vials, stoppers, seals and miscellaneous production equipment), formulation and filtration (the weighing, mixing and sterile filtration of the product), filling (the aseptic fill of vials), lyophilization (freeze-drying of the vials), sealing (stoppering of the vials and application of the aluminum overseals), inspection, labeling and packaging.

The MPBR section format should be consistent and should contain basic information, no matter what the product or process. The first page of each section should contain a bill of materials that lists the raw materials and components required to perform the processing event outlined in the section. It can easily be formatted in chart form.

In the bill of materials, each item should be listed with its part number, and a space should be provided to record the associated receiving code. List the actual quantity of each item required for processing, and provide a space to record the actual quantity of each item used in the process. (Note: In the formulation or assembly section, the quantity of raw material required per dosage unit must also be specified.) Provide spaces for the signatures of production and QC personnel who verify that those items were, in fact, the ones received and used in processing.

Each MPBR section should also contain an accountability section that lists the items from the bill of materials and provides space to record the exact amounts received for use, the amounts used, discarded, returned to storage, and so on. The percentage gain or loss should be calculated and an acceptance criteria established; signatures should verify that the calculations and entries are correct.

Who Writes an MPBR, and When?

Production and QC personnel should write the MPBR. It should be written as soon as the process has been well defined by the research scientist or when scale-up of the process is required to produce materials for analytical or animal testing programs. Although these early batch records may be incomplete in terms of GMP compliance, they provide an accurate record of changes that occur in processing during product and process development. MPBRs must comply fully with GMP requirements when used to document the production of product for clinical trials.

MPBR Language

A batch record is the document wherein the chemical or biological processes of drug manufacturing (developed in a research laboratory) are merged with the regulatory requirements of the FDA. Facilitate this often-difficult union of disparate disciplines with a well-written batch record that ensures the fundamentals of GMP compliance while providing a convenient, practical and efficient set of instructions for the line worker.

The MPBR should be written, therefore, both for the individuals who routinely perform the manufacturing operations and for the reviewing authorities. The MPBR must either clearly describe the activities that it directs or refer to documents, such as standard operating procedures, that describe those activities in greater detail.

The language of an MPBR should be similar to that of an SOP. Use active verbs such as "add this," "measure that," "adjust pH" and "record temperature." Specify the limits or range of acceptable values or observations whenever possible. Examples of language were discussed in the chapter on SOPs.

MPBR Content

In addition to providing step-by-step instructions on how to manufacture a product, the MPBR provides the opportunity to document that the production events occurred as directed. Consequently, it provides fill-in-the-blank spaces for recording technical information, observations and signatures.

The narrative portion of the MPBR should describe the manufacturing events chronologically. When one reviews a partially completed PBR, spaces and signatures that appear above that chronological point in processing should be filled in completely, and no signatures or blanks should be filled in below that point.

There should be signature spaces for documenting that processing events were completed as directed. These spaces should be completed by the production personnel who perform the event. If the event is critical—that is, if a deviation in that event could adversely affect the quality of the process or product—a signature space for either a second production operator or a QC technician is necessary.

Processing criteria should be stated on the MPBR with limits of acceptance, when appropriate. For example, "Adjust and record the pH of the solution to 6.0 +/- 0.2 with 1N Sodium Hydroxide (PN 1777)."

Other information that should appear in all MPBRs includes reference numbers that can assist in the complete traceability of materials or processes used (such as the location and/or room number of the area in which the processing occurs), date and time entries to document when processing begins and ends (as well as any time limitations on that processing) and identification of major equipment used in processing (such as sterilizers, lyophilizers, cell culture equipment, fermenters, filling machines, filtration hardware and mix tanks). All MPBRs should also contain sampling event data, including sample size and sampling instructions, when appropriate; cycle numbers from processing equipment such as sterilizers, lyophilizers and fermenters; processing parameters and/or calibration values such as temperature, pH, viscosity, pressure, vacuum, bubble points, CO_2, O_2, and glucose levels; calculations (fill-in-the-blank formulas); and exact copies of labels or printed materials used during labeling and packaging events. (Figures 5.1–4)

While designing an MPBR, review 21 CFR, Subpart F, Production and Process Controls, Section 211.100-211.115 to ensure that the process controls cited in these regulations have been included in the MPBR.

Cell-Derived, Bulk Production Batch Records

To extend the example provided, consider the production of a bulk protein derived from continuous animal cell culture. An MPBR would consist of sections on inoculation, culture monitoring and harvest, product purification, packaging, labeling, storage, culture termination and equipment decontamination.

In addition, there would be either batch record sections or forms to support the component preparation prior to inoculation. These major preliminary preparation events include media preparation, equipment preparation (cleaning, assembly, sterilization, operational qualifications, and so on) and cell line scale-up for inoculum preparation.

Whatever the format, the design of the MPBR and supporting documentation must

Our Laboratories, Inc.

MASTER BATCH RECORD: PRO10

Page 1 of 10
Edition # 01

SECTION A: COMPONENT PREPARATION

Product: Protein 33
    10,000 units/vial

Product Part # 4152

Lot #_____

PBR Approvals: _____ / _____

Theoretical Yield =
    20,000 vials

_____ / _____

MASTER PRODUCTION BATCH RECORD APPROVAL SIGNATURES

_____ _____ _____ _____ _____ _____
Production Date Quality Control Date Quality Assur. Date

BILL OF MATERIALS

Part #	Description	RC#	Quantity required	Quantity received	Prod.	QC Signatures
2111	vial, 10 mL 20mm		22,000			
2267	stopper, 20 mm gray		22,000			
2377	seal, 20 mm red		20,000			

ACCOUNTABILITY

Item	A = Qty received	B = Qty sterilized	C = Discards	D = Qty Ret'd to storage	% gain/loss
2111					
2267					
2377					

Calculation: A/B+C+D = % gain or loss; _____ / _____ = _____ %

Acceptance criteria = ± 5%

Calculated by _____ Verified by _____

(Figure 5.1)

Our Laboratories, Inc.
MASTER BATCH RECORD: PRO10

Page 2 of 10
Edition # 01

SECTION A: COMPONENT PREPARATION

Product: Protein 33
10,000 units/vial

PBR Approvals: _____ / _____

_____ / _____

Product Part # 4152

Lot # _____
Theoretical Yield =
20,00 vials

VIAL PREPARATION

1) Preclear the area according to SOP 501.   QC _____

2) Move vials into the area according to SOP 510. Verify that these are the vials cited on the Bill of Materials; verify that vials have been properly labeled and released. Enter event in room usage log.   PR _____

3) Record quantity received and RC#s on p.1.   PR _____

4) Load vials into washer and enter event in equipment usage log.   PR _____

5) Wash vials according to SOP 511.   PR _____

    Equipment ID # _____   Washer cycle mode _____

    Washer cycle # _____   Date/Time _____

6) Clean stainless steel trays for sterilization according to SOP 512 and prepare them for vial loading.   PR _____

(Figure 5.2)

Our Laboratories, Inc.

MASTER BATCH RECORD: PRO10

Page 3 of 10
Edition # 01

SECTION A: COMPONENT PREPARATION

Product: Protein 33
    10,000 units/vial

Product Part # 4152

Lot # _____

PBR Approvals: _____ / _____

Theoretical Yield =
    20,000 vials

_____ / _____

---

7) Load clean vials into trays according to SOP 512. Weigh one filled tray and record weight.

    one tray = ____ kg

PR _____

8) Load sterilizer chamber with filled trays according to SOP 513. Label trays and complete sterilizer equipment usage log:

PR _____

9) Calculate total weight of sterilizer load:
(wt. one tray) x (# of trays)

_____ x _____ = _____

PR _____

Minimum validated load = NLT 17 kg
Maximum validated load = NMT 560 kg

QC _____

10) Ensure that trays and RTDs are placed properly throughout the unit and that the loading configuration is correct according to SOP 513. Record sterilizer information on form 5/100. Load chart paper into recorder and write in the cycle number, date and initials.

    Equipment ID # _____

    Cycle # _____ Cycle mode _____

PR _____

QC _____

(Figure 5.3)

Our Laboratories, Inc.

MASTER BATCH RECORD: PRO10

Page 4 of 10
Edition # 01

SECTION A: COMPONENT PREPARATION

Product: Protein 33
      10,000 units/vial

Product Part # 4152

Lot # _____

PBR Approvals: _____ / _____

Theoretical Yield = 20,000 vials

_____ / _____

11) Start sterilizer.   PR _____

12) Unload sterilizer on clean side. Ensure that materials are properly labeled with part #, RC#, cycle #.   PR _____

13) Remove recording chart. Check against Master acceptable cycle charts and give to QC for review and approval.   PR _____

14) Clean sterilizer of broken glass or any other debris.   PR _____

15) Complete accountability section on page 1.   PR _____

> **NOTE:** A complete MPBR continues and documents stopper and seal preparation events.

(Figure 5.4)

ensure complete accountability and traceability of materials, equipment and personnel involved in the manufacture of product.

MPBR Numbers, Edition Numbers

Every MPBR is an official document, just like an SOP, and should have a number assigned to it to control its issue and change. In the example provided, the MPBR number is PRO 10. The format of an MPBR number generally provides information about the product and its configuration. If, for example, the batch record in Figures 5.1–4 were written for the production of a product containing 15,000 units/vial of Protein 33 instead of 10,000 units/vial of Protein 33, the MPBR number might be PRO 15 instead of PRO 10.

As discussed in a previous chapter, every controlled document must have an edition number. The edition number must be changed every time the document is revised, and a history of that change must be maintained. The edition number in Figures 5.1–4 is 01.

Product Part Numbers, Production/Batch Lot Numbers

Every MPBR must list the part number of the product produced. These product part numbers are assigned like any other part number; they have accompanying specifications that indicate all the characteristics of the final product and any testing required to release that product for further processing (if it is an intermediate product) or to the market (if it is a final product).

While the product part number indicates exactly what product is manufactured with the MPBR, the production lot/batch number indicates which production event—of many identical production events—created that batch of product. The assignment and use of these lot/batch numbers is discussed in the chapter on lot numbers.

Because a production lot/batch number is designed to distinguish between many production events of identical product, it is a time and/or location instructive number. Lot/batch numbers, therefore, usually contain reference to the month and year of manufacture. They sometimes contain reference to the site of manufacture (if there is more than one approved location for manufacture), and if appropriate, it can reference a certain production line within a facility (if more than one production line or machine is validated to perform the processing). Specific examples of production lot/batch numbers are presented in chapter 3.

The Code of Federal Regulations defines these numbers as follows: "Batch number or lot number means any distinctive combination of letters, numbers, or symbols from which the complete history of the manufacture, processing, packing, holding, and distribution of a batch or lot of drug product . . . can be determined."

MPBR Approval, Control and Use

As already mentioned, at least two company officials—one from the QA department and one from production—must approve each edition of an MPBR. These approvals must be documented with full signatures and dates.

Keep original MPBRs in a secure location. Official copies of the MPBRs are made only when a production order is scheduled; these official copies are called production batch records (PBRs).

When the PBR is issued, the lot number of the batch is filled in, and each page of each section of the batch record is marked or stamped by QC and/or production to indicate that it is an official copy of the MPBR. After the batch production has been completed, it is convenient to use a red stamp to mark spaces for approval signature(s) and the date to identify this document as an original. Alternatively, PBRs can be copied onto colored paper to distinguish them from MPBRs. Approval signatures, however, must still be applied.

Operational Suggestions

All of the policies and commitments concerning the creation, approval, use, and control of MPBRs and PBRs must be documented in SOPs. These documents must be included in the change control programs, and a history of change must be available for review.

PBRs must be followed during production events, and the information and signature blocks must be completed as production proceeds. It is an embarrassing moment—which many a pharmaceutical manager has experienced—when an FDA inspector asks to see a PBR for a batch currently in progress, and the PBR is only partially completed.

The Product Record

The product record (a term not specifically cited in the CFR) includes all of the documentation that supports the production and control of a single batch of product. The product record includes the PBR, QC records, sterilization charts, move tickets and any other critical batch-associated documentation. The product record files must contain original documentation. These records should be organized to facilitate both client and regulatory audits.

When a PBR has been completed, the production manager should collect all associated and supporting documentation and review it for completeness and accuracy. Supporting documentation includes sterilization charts, lyophilization charts, environmental monitoring records, move tickets, accountability records and inspection reports. This review process can be documented on a summary sheet that lists the forms or documents required for a complete batch and verifies—with signatures—that the documents are available and complete. Any deviations that may have occurred are brought forward for discussion or investigation.

When the product has been tested and approved for release by QC, the QC records are assembled. These records include raw material analysis, in-process testing results, final testing, environmental and microbiological monitoring, pre-clearance activities, inspection reports, and so on. The records are then reviewed for completeness and accuracy, as mentioned above. Again, any deviations are brought forward for discussion or investigation.

Final release of product occurs only when the entire product record has been reviewed and found to be acceptable by production, QC and QA. This review and approval is mandated in 21 CFR 211.192 and must be documented in writing. The product record is then filed in a secure location.

It is helpful to remember that the product record contains evidence that can be used in a courtroom, and as such can be used against the company to prove wrongdoing. Conversely, it can be used to support the company by documenting acceptable operations—the difference is often a matter of documentation system design.

6

Equipment Installation and Identification

THE MAINTENANCE (OR ENGINEERING) department is a critical first link in ensuring the quality of all operations in a GMP production facility. Failing to acknowledge the importance of GMP requirements in maintenance can quickly compromise all other operations in the facility.

The maintenance department itself is, in fact, a "bulk chemical" production facility. It continually manufactures critical raw materials, such as water for injection, clean steam, clean air, compressed air and nitrogen. Clearly, raw materials of poor or inconsistent quality significantly affect production department operations and, eventually, the final product.

The equipment and the processes that the equipment supports within the maintenance department must therefore be controlled and documented to meet GMP requirements. This control is best achieved by validating equipment installation, operation and performance and then establishing ongoing systems to document equipment repair and maintenance activities.[4]

Installation Qualification

An equipment installation qualification (IQ) evaluates the assembly and installation of critical processing equipment to ensure its safe operation. During an initial IQ, it is also necessary to confirm that the equipment ordered is what was in fact received, and that it is capable of supporting its intended function (consult 21 CFR 211.63 and 211.65).

IQs should be performed every time a unit is re-certified or validated. Annual qualification ensures that no significant changes have occurred in the system or, if changes have occurred, that they have been properly documented and controlled. This is a GMP requirement. This periodic review should also include an audit of the repair and maintenance history of the unit as documented through work orders and preventive maintenance program documents.

There is no established format for an IQ document, but some content guidelines should be considered. These include equipment identification, utility requirements, safety features and any critical installation specifications.

Equipment identification. Include the in-house identification number (discussed below), location of the equipment, its manufacturer and supplier, its manufacturer's model and serial numbers, its dimensions and capacity, and when appropriate the accounting asset number and/or blueprint drawing number references.

Utility requirements. When listing utility requirements—including services such as water, electric power, gas, compressed air, steam, nitrogen, and drain and exhaust lines—indicate the quality of the feed utilities and any quantity or volume requirements. For example, an autoclave requires not just steam, but clean steam, and also has a minimum steam pressure requirement. Similarly, autoclave steam feed lines may have minimum diameter requirements and material requirements of 316 stainless steel. The IQ should list these requirements.

Safety features. Describe equipment safety features, including any requirements for relief or pressure-reducing valves and alarms. The IQ should indicate when an alarm will sound or initiate an automatic shutdown.

Installation specifications. The IQ should include any other critical installation requirements, such as those for ventilation, discharge-line sizing, temperature, alignments and sound proofing. The maintenance engineer should consider what could damage equipment, shut it down unnecessarily, compromise the quality of the process it performs or threaten the safety of the work environment. Then the engineer should determine what installation requirements could eliminate these potential problems or minimize their impact, should they occur.

Preoperational cleaning and safety checks. These can be performed either as a part of the IQ or as a part of the operational qualification discussed later. Preoperational checks include tasks such as cleaning and/or passivation of holding tanks and distribution lines, pressure tests of closed systems and sanitization or sterilization of equipment. Critical preoperational checks must be performed; their location in the validation format is an individual company policy decision.

Functional Descriptions

Food and Drug Administration inspectors commonly ask for "a description of how this equipment or system works." It is useful, therefore, for everyone who works on or with a piece of equipment to understand how it operates and how it might function in relation to other utility or production systems. This information can be included in a narrative description of the unit that details the basic principles of operation, how the operation is controlled or monitored and how operations might affect other equipment. This description can be located in the IQ document, in the introductory section of the operating SOP or in a master validation protocol.

IQ Documentation

When equipment is installed for the first time, maintenance writes, signs and dates the original IQ document. IQ documents must be signed and dated. They list the installation criteria, reference "as built" drawings and declare that the equipment is properly installed.

When the IQ is complete, equipment can be released for further evaluations, such as operational and performance qualifications. If the facility requires annual validations, annual IQs should be performed as a first step in validation.

When a maintenance documentation system has been properly designed and implemented, this annual event should require a minimum of effort. When the previous IQ document is reviewed, a new document can be issued and signed if no changes have occurred, or a revised document can be issued and signed with changes indicated and justified. The annual IQ event also offers the maintenance engineer an opportunity to review all repair and maintenance records on that equipment and to revise preventive maintenance schedules and activities accordingly.

Equipment SOPs

It is convenient to write SOPs for equipment operation when drafting the initial IQ. SOPs should be written for the start-up, alarm reset and shut-down of all major equipment systems, such as water-for-injection systems, clean steam generators and compressed air. These SOPs should contain routine maintenance operations, such as blowdown of boilers, filter changes and regeneration of deionized water system resin beds.

Equipment cleaning and maintenance SOP requirements are cited in 21 CFR 211.67, which discusses the need for written procedures that describe the "cleaning/ sanitization/sterilization of," "assembly/inspection of," "operation of" and potentially the "disassembly of" production equipment. Although the production department often writes these procedures, maintenance should have direct input into—if not responsibility for— their content. Clearly, maintenance input depends on the critical nature of the equipment and/or on the critical nature of the production process it supports.

Routine maintenance tasks, which I will discuss in the next chapter on preventive maintenance programs, must also be documented in procedures. These routine tasks can be drafted during the initial equipment installation phase and finalized after the initial validation of systems. It is impossible and unnecessary to write procedures for every event that a mechanic or engineer encounters during a routine work day. But critical operations on critical equipment (discussed below) must be performed reliably and consistently. An SOP facilitates these activities.

Employee training programs within the maintenance department and from outside vendors are important for ensuring that unusual and unexpected repairs occur in a proper manner. It is common practice for individual mechanics within the department to become experts on certain systems or equipment.

Equipment Parts and Materials Specifications

During the initial IQ and design of preventive maintenance programs, maintenance must determine which parts and materials associated with equipment operation require quality control. For example, in-line filters in a purified water system must receive part numbers and receiving codes and be subjected to QC inspection and/or testing procedures before they can be used in production.

Also, the feed water for the purified water system must meet specifications—QC must assign specifications to the feed water and monitor it routinely. Other materials to be considered for quality control inspection and testing include lubricants, compressed gases, deionizing resins, water softening agents, boiler additives, vent filters, cleaning solvents and deaeration chemicals.

Consult with QC to determine which items will require inspection and/or testing before their release to the maintenance department. When writing these materials specifications, consider safety issues in addition to product quality concerns.

Equipment ID Number Format

Equipment identification (ID) numbers are not strictly required by GMPs. Their use, however, facilitates GMP documentation requirements, cited in 21 CFR 211.105. Some of these GMP documentation requirements include equipment installation qualification records; production batch record citations for critical equipment used in processing; and maintenance/repair records for equipment, including a preventive maintenance program and any equipment change control practices.

Equipment ID numbers are simple numbers or alphanumerics, but their format should be different from other numbering systems used in the facility. It is helpful if the numbers contain some information about the equipment they identify.

Before designing an equipment ID numbering system, consider the appropriate parameters for the number format. An example is an equipment ID numbering system with this format: 21C/301/K. Here, the first two numbers indicate the type of system the equipment supports: "category 20" equipment supports water systems (21 = water-for-injection still, 22 = deionized, purified water system, 23 = reverse-osmosis system); "category 30" equipment supports steam systems (31–33 = boilers, 34 = clean steam generator, 35–37 = autoclaves, 38 = steam-in-place system); "category 40–50" equipment supports conditioned or controlled air systems (41 = compressed air, 42 = HEPA filtered area, 47 = laminar flow hoods, 54 = nitrogen system, 55 = incubators, 56 =

ovens, 57 = refrigerators, and 58 = freezers).

The letter "C" indicates that this is a critical piece of equipment. In this example, the number 21C/301/K identifies a recirculating pump on a water-for-injection line, and the "C" means that before one can proceed with repair of this equipment, the QC and production departments must be notified. Notification is necessary because pump repair could compromise the quality of the water for injection. In a worst-case scenario, revalidation of the water system may be required. An "R" in this position might indicate that this equipment is a part of the preventive maintenance program and that repairs must be documented according to SOP. An "X" could indicate that the equipment is serviced by an outside contractor. The lettering system can be developed to meet the needs of the individual facility. Its purpose is to inform the maintenance engineer quickly and conveniently what level of concern and/or documentation is required.

So far this piece of equipment has been identified as belonging to a particular system, "21" (water-for-injection), and designated as a critical piece of equipment with the letter "C." The next three digits in this equipment ID number, "301," specify a unique number for the pump. The storage tank in the same system might be labeled 23C/402/K.

Finally, the "K" is offered as a location indicator. It might refer to one of several buildings, or it might refer to a department within a facility. Similarly, a room number designation could be added.

Equipment ID Number Assignment

All critical equipment that comes in direct contact with the product during production should be assigned an equipment ID number. Whenever that equipment is used in critical processing events, this number is recorded in the production batch record.

In addition, all equipment that requires routine calibration, maintenance evaluations or repairs should be identified with a number. Again, this number simply facilitates record keeping. Whatever the guidelines established for assigning equipment ID numbers, the system should be described in an SOP. A list of number assignments should be maintained in the maintenance department or by QA, and the next chapter discusses the supporting files that should be established.

When the number has been assigned, use it to label the equipment. Labels can range from plastic adhesive labels to metal tags. Ensure that the number is easily visible during equipment operation and that it will not deteriorate during equipment cleanup or use.

Change control is fundamental to the maintenance program and must be documented. When there is a major change in a system, it may be appropriate to assign a new equipment ID number and retire the old number; or it may be appropriate simply to write and perform a new installation qualification and/or validation.

Clearly, in the example of an equipment ID number provided earlier, relocating the equipment could change the number from 21C/301/K to, for example, 23C/301/V.

Our Laboratories, Inc. Page 1 of 2

Master Equipment Card

Equipment Identification Information:

Equipment: _____ **Equipment ID #** _____

Manufacturer/Supplier: _____

Model # _____ ; **Serial #** _____

Asset # _____ ; **Purchase order #** _____

Date of installation: _____

Equipment is a component of system: _____

Capacity/HP/RPM description: _____

Size/dimensions/weight description: _____

Output specifications: _____

   Output piping material requirements: _____

   Output piping size requirements: _____

Manufacturer/Supplier contact: _____

_____

_____

     phone   _____

     FAX   _____

Outside calibration or repair services: _____

_____

_____

     phone   _____

     FAX   _____

(Figure 6.1)

Our Laboratories, Inc.

Page 2 of 2

Spare Parts/Supplies*	Catalog #	Vendor	In-House Storage #

*Includes items such as oil, lubricants, filters, gaskets, and seals

Information and Documentation Resources:

    Validation protocol? ( ) No ( ) Yes _____

    Pertinent SOP #s _____

    Drawing reference #s _____

    IQ performed: ( ) Yes ( ) No (if no, complete next section)

Utility requirements: _____

_____

Safety features/Set-points: _____

_____

Critical installation requirements: _____

_____

(Figure 6.2)

Similarly, moving the preventive maintenance program to an outside contractor might change the number to 23X/301/K.

Equipment History Files

When equipment has been installed and assigned an ID number, an equipment history file can be created. These files contain all reference information for the equipment as well as a history of change, repair, and maintenance. They are equivalent to the part number specification files in the QC department. Equipment history files should be kept in a secure but accessible location within the maintenance department.

Master Equipment Cards

The first section of an equipment history file contains the master equipment card, which shows the equipment ID number, date of installation, any in-house asset numbers and the specific location of the equipment. Often the identifying information from the IQ document, if one exists for the equipment, also appears on the master equipment card. However, because many pieces of equipment in a facility require no formal installation qualification, the master equipment card provides a common source of information on the unit. (Figures 6.1 and 6.2)

Information that will facilitate the repair and maintenance of equipment and systems should also appear on the master equipment card. This includes references to and/or copies of SOPs on the use of the equipment, preventive maintenance program references and calibration procedures; the equipment manufacturer and/or supplier, address, telephone number and contact person; spare parts vendors, catalog numbers, telephone numbers and alternate vendors; any storage reference numbers for the in-house spare parts warehouse; and any outside calibration services or outside repair services (the availability of these contractors should be listed, even if they are not routinely used).

7

Equipment Monitoring, Repair and Preventive Maintenance

ONCE EQUIPMENT HAS BEEN PROPERLY installed and validated, it must be maintained to ensure that it continues to meet validation criteria throughout its life cycle. The commitment to design maintenance programs and their documentation systems can be met in a number of ways. General requirements are described in the Code of Federal Regulations; the rest is common sense.

Good Manufacturing Practices cite two general requirements for the routine cleaning, inspection, use and maintenance of equipment. To comply with 21 CFR 211.67, an equipment maintenance program must be established with implementation schedules and written instructions. This program must be followed, maintenance activities must be documented and these records must be available for review. The content and function of equipment cleaning and use logs is described in 21 CFR 211.182.

The basic documentation tools of an equipment maintenance program are equipment log books, maintenance and repair work orders, and preventive maintenance schedules and forms. These documents have no set format, but content guidelines must be considered when designing a system.

Log Books

A log is a chronological record of all equipment-related activities; it provides easy access to information about equipment status at any moment in time. There will be log books for equipment operation as well as room usage, cleaning and use.

Equipment. Equipment log books should be available for all major equipment and systems in a GMP production facility—including boilers, water-for-injection (WFI) stills, purified water systems, packaging equipment, cell culturing machines, fermenters, fill lines, lyophilizers and sterilizers.

Room usage. In some cases a room usage log book replaces an equipment log book. When a controlled area of the facility is dedicated to a certain activity and supported by specific equipment, room use and equipment use become synonymous. For example, a vial-labeling operation may always occur in a dedicated room. There, one log book can be both a room and equipment log for documenting room cleaning, equipment maintenance and batch labeling activities.

Room cleaning and use. Some rooms with no major equipment installations require room cleaning and use log books. These rooms are usually controlled access areas—often clean rooms. A room used for master cell bank work, for example, requires strict control over cleaning and usage. No two cell lines may be in the room at the same time, and cleaning and monitoring must occur between cell-handling events. All of these activities are recorded in the room cleaning and use log book. Before technicians enter a room to work with a new cell line, they must record this event in the log book. If no cleaning event is entered into the log book before the technician enters, a GMP violation occurs.

Format. Minimum log book entries include date, time, technician and event. Other appropriate entries depend on the equipment and the system. A boiler log book, for example, may list an established number of routine tasks, such as start-up, top blowdown, bottom blowdown, shutdown and sampling. The list permits the mechanic to simply check off, sign and date the event. Similarly, when a technician monitors equipment operating parameters such as line pressures, flow rates, temperature and conductivity, the equipment log book is a convenient place to record data. Log books are also appropriate for analytical equipment used for several different purposes by several individuals or departments—HPLC units, for example. Equipment calibration, cleaning, column changeover, reference profiles and sample runs are examples of appropriate log book entries.

Production equipment log books must also list the batch numbers of product processed by the equipment. This requirement is specifically outlined in 21 CFR 211.182:

"A written record of major equipment cleaning, maintenance . . . and use shall be included in individual equipment logs that show the date, time, product, and lot number of each batch processed. If equipment is dedicated to manufacture of one product, then individual equipment logs are not required, provided that lots or batches of such product follow in numerical order and are manufactured in numerical sequence. In cases where dedicated equipment is employed, the records of cleaning, maintenance and use shall be part of the batch record. The persons performing and double-checking the cleaning and maintenance shall date and sign or initial the log indicating that the work was performed. Entries in the log shall be in chronological order."

The log should be a bound book with numbered pages and column headings predrawn for all categories. The log book must be stored on or near the equipment it documents. Plastic file holders are convenient for attaching the books to the machinery or a nearby wall.

The documentation department should issue log books in a controlled manner. When a book is filled, it should be returned to documentation for storage.

Equipment Work Orders

Routine equipment operation creates a continuing need for unexpected or emergency repairs. The resulting repair work, as well as routine maintenance work, can be documented with a work order. The sample work order provided here is a simple, one-page form bound with four identical copies. It is prenumbered and divided into three sections. (Figure 7.1)

Section 1. The individual or department requesting the services of the maintenance staff completes the first section, briefly describing the problem. The requester indicates whether the problem is an emergency and whether repair is likely to affect a critical production event. If the repair could disrupt a critical utility or stop production, then both QA/QC and production must be notified when the repair occurs and must sign the work order to acknowledge notification. When the request is completed, the requester keeps one copy of the work order and sends the remaining copies to maintenance. Maintenance mechanics can contact this person for a full description of the problem.

Section 2. Maintenance performs the work and completes the second section, providing a brief description of the repair work. Attachments are acceptable if the work is extensive.

Section 3. The work order's final section documents completion of the work by maintenance and approval and release of the equipment into routine operation. If further testing must be performed on the equipment before its release, this option can be checked.

When the work order has been completed, maintenance returns one copy to the requester and one copy to QA. It retains one copy for a chronological work-order file and, if a master equipment file for the equipment repaired exists, one copy for that file. The format and content of master equipment files are described in greater detail later.

Preventive Maintenance, Calibration Programs

A preventive maintenance (PM) program is cited as a GMP requirement in 21 CFR 211.67. Beyond that, ensuring the reliable and consistent performance of equipment is simply good sense and good maintenance practice in any production facility.[4] There is no established format for a PM program, yet all industries offer effective examples, some of which are available as software programs.

Our Laboratories, Inc. WORK ORDER # ____ [preprinted] ___

Maintenance Work Order

MAINTENANCE REQUEST: Date of request _____ Department _____

Location/Description of repair activity (equipment ID#/room):

_____
_____
_____

( ) Emergency   ( ) ASAP need by _____   ( ) need by _____
                                    (date)   (date)

Will this repair affect a critical production process?

( ) Yes   ( ) No (if yes, QC and Production must sign this request)

_____   _____   _____   _____
Requester   (date)   **Quality Control**   (date)

_____   _____
Production   (date)

- -

Request received by Maintenance on _____ (date)

Summary of repair work: _____

_____
_____
_____
_____

( ) Attachments accompany this work order

Repair completed on: _____ (date)   Mechanic _____

- -

( ) The unit or system is available for routine use.
( ) The unit or system is available for QC testing.

_____   _____   _____   _____
Maintenance   (date)   **Quality Control**   (date)

_____   _____
Production   (date)

(Figure 7.1)

Whatever their format, all PM programs have a few basic elements in common. First, each piece of equipment and each system must be evaluated to determine what maintenance and calibration tasks should be performed periodically to help ensure reliable and consistent operation. Appropriate and important PM tasks can be determined from vendor or manufacturer equipment manuals, through experience, and through trade and industry association guidelines from organizations such as the Association for the Advancement of Medical Instrumentation, the American National Standards Institute, the American Society of Mechanical Engineers, the International Society of Pharmaceutical Engineers, the Parenteral Drug Association, the Health Industry Manufacturers Association and the Pharmaceutical Manufacturers Association.

PM documentation. When PM tasks have been determined, they are listed—along with the frequency of performance, the methods of evaluation (sometimes written in SOPs), and any acceptance criteria—on an equipment preventive maintenance or calibration task list. The list can be filed in the master equipment files and in a separate PM program notebook.

Complex PM tasks must be performed according to written instructions, which may be in the form of an SOP or in a newly designed, more specific preventive maintenance procedures (PMPs) format. The performance of each PM task must be documented. This documentation can be on SOP-associated forms. An alternative is to format a PMP as a preventive maintenance record (PMR)—similar to a production batch record—that describes each task with step-by-step instructions and provides fill-in-the-blank spaces for data and signatures.

Whether one chooses to write SOPs and forms or a PMR, these documents must be approved and subjected to corporate change-control policies. Copies of the documents should be filed in the master equipment files as well as in a PM file.

Master PM checklist. A master PM checklist is compiled from the individual equipment PM task lists, detailing weekly PM tasks to be completed throughout the year for the entire production facility. The maintenance manager must design this master PM checklist to coincide with annual or semiannual equipment validation requirements and calibration schedules to ensure an even workload.

Work from the master PM checklist can be assigned in many ways—ideally by giving each mechanic and engineer a monthly PM assignment list with copies of the appropriate blank PMRs. Completed records are returned to the maintenance manager for review and approval, and the task is checked off the master PM checklist. Finally, these PMRs are filed in the master equipment files and, sometimes, chronologically in the PM program file. All of these manipulations can be accomplished with the appropriate computer programming. Many preventive maintenance operations are administered in this manner.

Although a preventive maintenance task or routine maintenance activity may be properly documented in the PMR, the need remains for operators and technicians to know that those events have occurred. Calibration and preventive maintenance activity stickers can accomplish this. Apply stickers to equipment when operations are completed. Include the date of the activity and the next date that activity is scheduled to be performed.

Master Equipment Files

Master equipment files—also called equipment history files—have been referenced throughout this chapter. Master equipment files are equivalent to QC specification files in that they contain all pertinent information about a piece of equipment and a history of its repair, maintenance and use. Master equipment files contain items such as master equipment cards, copies of the IQ, "as built" drawings, manufacturers' manuals, copies of relevant SOPs, the equipment PM task list, copies of preventive maintenance procedures or PMRs, relevant completed work orders and any other related information or memos.

8

Designing GMP and Facility Qualification Master Protocols

PROTOCOLS ARE NARRATIVE DOCUMENTS that describe the general conduct of operations or activities in GMP facilities. Protocols contain company policies or commitments and cite more specific documents—such as standard operating procedures (SOPs)—in order to unify individual activities within or between departments. In effect, protocols describe the routine rituals of conduct in a facility that ensure consistent, reliable operations. Protocol commitments usually include company or departmental policy, standards and acceptance criteria.

Neither the Code of Federal Regulations nor Guideline documents specifically mention protocols, except in association with validation. Although the documents themselves are not required, the information and commitments they contain are.

This chapter presents protocols as documents that unify, direct and control activities in a GMP environment. Throughout inspections and audits by clients and regulators, protocol documents provide a clear, organized map of the often overwhelming sea of SOPs, specifications and product processing events.

Protocol Format

There is no established format for a protocol, but some basic information should be included. Each page of the document should contain the name of the company, the number and/or name of the protocol, an edition number and a page number. Each document must be approved and approval signatures must appear on at least one page of the protocol.

Other parts of a protocol include the purpose and/or introduction, references, scope, responsibility, acceptance criteria and documentation requirements.

Purpose and/or introduction. The purpose of the protocol presents an overview of principles/concerns addressed in the document.

Scope. This section describes activities to which the protocol does and does not apply. The scope can also be used to place this protocol in context with other protocols or procedures.

Responsibility. The person or department responsible for the conduct of the protocol is stated here. Often several departments are involved; if so, this can be presented in this section of the protocol.

References. This optional section can provide a source of additional information on the subject, allied SOP documents, CFR or Guideline references.

Protocol. The protocol itself can be formatted as required by the information it presents. Validation protocols, for example, have sections for preliminary operations, installation qualification, operational qualification and performance (process and product) qualification. Use a clear, logical format that facilitates its use and understanding.

Acceptance criteria. In validation protocols, the validation is considered complete only when acceptance criteria have been met. Similarly, protocols for certifying vendors, evaluating sterility testing apparatus or accepting an environmental control program should contain acceptance criteria commitments. State acceptance criteria as a range of values or as minimum and maximum acceptable values.

Documentation. Requirements should describe how events cited in the protocol are to be documented, reviewed and approved. This section may include reference to data collection forms, memos and certificates.

When to Write a Protocol

Protocols present a company's "big picture" commitments or policy. Consequently, they are ideally written before SOP writing begins. It is best to write a master validation protocol, for example, before writing individual validation protocols for a water-for-injection system, an autoclave or a lyophilizer.

The **master validation protocol** (MVP) describes the format and content of individual validation protocols. It provides guidelines for determining the amount and type of testing to be performed for an individual validation event. It categorizes major equipment, systems or processes according to the rigor of testing required. The MVP describes validation scheduling requirements, the order of testing and the frequency of validation (recertification).

With a short timeline for facility approval, in real life most protocol documents are written when there is time or when someone thinks of it. Writing the "same things" in several SOPs may signal the need for a protocol that declares these "same things" about those SOPs and refers to the individual documents.

Protocol and edition numbers. Every protocol should be identified and controlled with two numbers. Protocols should have identifying numbers and edition numbers. Because there are generally fewer protocols than SOPs, a simple identification number—P26, for example—is usually adequate. Number assignment, approval, issue, distribution, and control of protocols should mimic the systems for SOPs. Edition number or revision level numbers of 01, 02 or A, B, C will serve to identify and track document change.

Appropriate topics. Validation is a common protocol topic because validation protocols are required for GMP compliance. Other candidate topics include product stability testing, standards for packaging and shipping product, environmental control programs, warehousing practices, vendor certification, departmental responsibilities (for the quality control department, for example), GMP training programs, environmental bioburden testing programs, sampling procedures and standards, as well as documentation basics. Protocols might also be guidelines for the release of finished goods, for quality control evaluations during product manufacture, for facility visitors, for release of information from the facility, for the use of consultants and outside contractors and for handling product complaints and returns.

Facility Qualification Master Protocol

The facility qualification master protocol (FQMP) outlines management's commitment to meet GMP requirements. It describes the tasks to be performed and the acceptance criteria that must be met before the facility or product line is considered available for Food and Drug Administration inspection and/or GMP manufacturing events. The FQMP describes the systems that support GMP in the manufacturing facility.

When the FQMP is complete and the data that support it have been collected and approved, it can be compiled into a master facility qualification package or document. This package should be kept in the same location as completed validation protocol packages or documents. The FQMP is often the first document that an FDA inspector asks for, and it should describe the company and the manufacturing site in a manner that facilitates a GMP inspection.

An FQMP contains the following elements:

Purpose. The statement of purpose describes or summarizes the qualification of the facility—for example, Our Laboratories, Inc., in Tofte, MN. It includes a statement such as, "When the programs described in this protocol are complete and acceptable, the facility will be available for FDA inspection and approval."

Scope. The statement of scope can read something like, "This protocol directs the initial start-up and validation of the Our Laboratories facility in Tofte and its associated product lines." (Separate facility qualification master protocols are prepared for each Our Laboratories manufacturing site.)

Responsibility. The section on responsibility includes a statement such as, "These activities are directed by the vice president of operations. Final compilation and audit of procedures, protocols, and data that support this qualification is the responsibility of the quality assurance department."

Protocol. Format the protocol as dictated by the information you are presenting. An FQMP includes sections on the business, departmental organization, building and facilities, the master validation protocol, the master GMP documentation system design protocol and GMP monitoring and control systems.

The **business section** briefly describes Our Laboratories as a business and provides a brief history of the company's origin. Here, identify the company as private or public and describe its relationship to other companies or organizations (for example, it may be a subsidiary or division of a larger company, or be involved in a strategic partnership). Cite any changes to the corporation's name that may have been made over the years. If the facility is new, explain any current need for expansion, what portions of the business this facility will support and the product lines—and associated processing—to be produced at this facility, noting any contract products. Also discuss future plans for this facility.

The section on **departmental organization** describes the departments in this facility and provides a diagram of reporting structures with position titles. Refer to any SOPs on departmental responsibilities and/or describe those responsibilities in a paragraph.

The **building and facilities** section includes a floor plan of the building that indicates room numbers and departmental areas. It provides narrative descriptions of the major utility systems in the facility that support GMP manufacturing—for example, water-for-injection, purified water, house steam, pure steam, CIP/SIP systems, HVAC, class 100 clean rooms, sterile compressed air and waste treatment.

The same section also provides brief narrative descriptions of the major product process equipment and/or systems in the facility, including those for fermentation, cell culture, purification, formulation, aseptic fill and packaging. Describe the flow of raw materials through the facility as well as personnel, product, waste and air. A floor plan marked with colors and arrows is a good way to demonstrate that the potential for product mix-ups and contamination events has been minimized. It is also a good idea to designate containment areas on the floor plan, indicating areas of coverage for separate air-handling units, air-flow patterns, pressure differentials, areas with filtered exhaust and areas with once-through air systems.

In the **master validation protocol**, include a statement like "Our Laboratories facility in Tofte will be validated to meet GMP requirements. These commitments are detailed in the master validation protocol (Protocol P15), which describes the format, approval, review, distribution and control of validation protocols for equipment and utility systems, manufacturing processes, and analytical methods." This is also the place to provide a list of utility, equipment and method, as well as processing that occurs at the facility. Categorize each list according to the rigor of the evaluation; this categorization will be discussed later in this text. Also cite references to individual validation protocols for equipment, utilities, methods or processes listed.

The **master GMP documentation system design protocol** states that "the general principles of GMP documentation for the Our Laboratories facilities are presented in Protocol P17. Individual components of the documentation system such as part numbers, specifications, SOPs, lot numbers and work orders are presented in this document."

The section on **GMP monitoring and control systems** can begin by stating, "the programs and systems for monitoring and controlling operations within this GMP facility are outlined in Protocol P21, which discusses GMP personnel training, and programs for environmental control, material movement and control, production process control, quality control assessment, vendor certification, preventive maintenance, calibration and metrology, quality assurance audit, and final product monitoring and control."

Acceptance criteria. When the commitments in the facility qualification master protocol document have been completed and reviewed and approved by—in this case—the vice president of operations, approval is stated in writing. Management acknowledges that any major change in the company's business, facility construction, personnel or products could affect the GMP acceptability of this facility. It is appropriate to state here that "this protocol will be audited or reviewed as appropriate, and at least annually, to assess the impact of any changes."

Documentation requirements. Describe the final format and location of a completed FQMP.

The FQMP should be the first document drafted when initiating the start-up of a new facility, but it should also be the last document completed. Only when all of the commitments stated in this document are fulfilled is the facility open to FDA inspection.

9

Master Validation and Equipment Validation Protocols

VALIDATION MUST BE DOCUMENTED to ensure that a specific process, method or equipment system consistently produces a product that meets predetermined specifications and quality attributes.[6,7] A validation protocol is a written plan that describes how to conduct validation and how to measure the success of the process, method or equipment performance. It identifies acceptance criteria for raw materials, key processing variables, intermediates and final product.

The three major types of validation protocols cover equipment and utility systems, methods, and processes. Because each type of manufacturing environment has a unique mix of validation requirements, it is instructive to write a master validation protocol that describes the commitment to validation in the context of your facility, your processing lines and your products.

The master validation protocol documents the company's commitment to validation. It describes the types of validation to be performed and the format and content of the various validation protocols. It lists the equipment, utilities, methods and processes that will be validated.

Although many critical processes, methods and equipment systems require a full validation, many others require less rigorous analysis. It is appropriate, therefore, for the master validation protocol to provide guidelines for deciding when validation is appropriate and to specify who will make the decision. The decision-making body is usually a validation committee consisting of management representatives from production, QC, QA, material handling and maintenance.

In addition, it is appropriate for the master validation protocol to address content issues that apply to all validation protocols, such as order of validation. It would be inappropriate, for example, to validate an autoclave before validating the clean steam system. Similarly, it would be inappropriate to validate an aseptic filling process before validating sterilizers or any analytical methods used to assess product quality.

A mechanism is needed for handling deviations that occur during validation; the protocol should establish time limits between installation, operational and performance qualification. It should also declare revalidation requirements and guidelines.

Method and process validation protocols will be discussed in the following chapters; this chapter will focus on equipment and utility system validations.

Validation protocols include three major types of evaluation: installation qualification (IQ), operational qualification (OQ) and performance qualification (PQ). A validation is complete only when all three qualification procedures have been completed, in order, and all acceptable criteria have been met.

Installation Qualification (IQ)

An installation qualification study should establish confidence that the equipment is properly and safely installed. To do so the structural installation must meet the manufacturer's suggested guidelines, and all utility support systems—electric, gas, steam, compressed air, water—must meet design limits and codes.

An IQ document can simply list the installation specification and requirement limits and provide a signature block for maintenance to ensure that the equipment requirements and specifications are met and will continue to be met. To be qualified, new equipment received must match equipment ordered. Any exceptions to original purchase specifications must be documented.

The three categories of equipment information to consider for an IQ document are identification information, utility requirements and safety features.

Equipment identification information includes any in-house equipment identification numbers, the manufacturer's model and serial numbers, equipment size, dimensions, weight, capacity and location (room number) in the facility. Include references to blueprint and drawing numbers, when appropriate. Much of this information also appears in master equipment cards, as discussed in a previous chapter.

Equipment utility requirements include services such as water, electric, gas, compressed air, steam, nitrogen, drain and exhaust lines. Indicate the quality of feed utilities and any quantity or volume requirements. Similarly, indicate any specific piping requirements such as pipe composition, diameter, filter requirements and reducers.

Equipment safety features include pressure relief valves and alarms and the settings at which they are designed to activate. Other potential safety requirements include room temperatures, discharge-line sizing, alignments and soundproofing.

Operational Qualification (OQ)

An equipment operational qualification study establishes that the equipment can operate within established limits and tolerances. This demonstration of basic equipment performance must involve all operations that will be used routinely in manufacturing.

The four categories of information to consider for the OQ document are calibration requirements, preoperational activities, operations and acceptance criteria.

Calibration requirements must be stated in the OQ. Indicate the parameters, how they will be measured or monitored, and what is considered an acceptable range or limit. Include all critical measuring and monitoring devices such as timers, pressure indicators, flow meters, temperature sensors and any chart recorders that document performance. If a sensor serves a controlling function within the unit, describe how this control is achieved and/or monitored throughout a cycle.

Preoperational activities must be declared, both those tasks unique to initial start-up and those appropriate for annual or periodic OQs. These activities include cleaning and sanitization of piping systems, passivation of stainless steel tanks and/or distribution lines, and software checks. In addition, it may be appropriate to check, for example, door gasket integrity, vibration of blowers and motors, door interlock functions, cycle set points and the performance of all heating elements. Describe each activity in detail in the OQ or reference it in an SOP.

Operations criteria should describe how to exercise all electromechanical options on the equipment and describe acceptance performance. An autoclave OQ, for example, would include a description of each cycle, directions for starting the cycle and acceptable results of cycle monitoring. Each of these exercises is performed with an empty chamber.

In addition, uniform heat distribution within a chamber is demonstrated during OQ with data collected from thermocouples distributed throughout the chamber. Three consecutive, successful empty-chamber cycles demonstrate uniformity. It is also instructive to include a fully-loaded chamber cycle to ensure that the distribution is not significantly affected by the presence of components in the sterilizer.

Data generated by the heat distribution studies are analyzed to determine the cold spot in the chamber. Traditionally the cold spot in a steam sterilizing autoclave is the drain; therefore, the controlling resistive temperature detector (RTD) for the cycle is appropriately located in the drain. If the heat distribution studies indicate a different cold spot, the controlling RTD should be moved to that cold spot before beginning heat penetration studies. The validation protocol should indicate how this determination will be made.

Because the concept of validation began with sterilizers, it is common to use them as examples. Sterilization systems are plentiful in drug and biologics manufacturing— not only steam sterilizers, hot air sterilizers, steam-in-place systems, sterile filtration, sterilization of lyophilization chambers, but also final product radiation and ethylene oxide procedures. Because OQ requirements are more rigorous for a sterilization process than for many other systems, it is appropriate to use this established format as a guideline for designing other equipment validation protocols.

Acceptance criteria for the OQ must be stated in the validation protocol. At the conclusion of this phase of testing, and before proceeding to the performance qualification, the data must be reviewed and a declaration written stating that results are acceptable.

Performance Qualification (PQ)

Once it has been established that the equipment is properly installed and functioning within specified operating parameters, it must be demonstrated that, if necessary, it can perform reliably and reproducibly under worst-case conditions.

Performance should be demonstrated with product under routine operating conditions. Worst-case performance for a piece of equipment can be demonstrated by assuring that all ideal processing parameters, defined during the OQ, are met with maximum loads or capacities.

For utility systems such as WFI or steam, the process qualification can demonstrate the production of acceptable-quality water or steam during three cold starts at all points in the system. Worst-case conditions for utility systems, however, are best evaluated over time with rigorous monitoring programs, as it is only when filters fail, regeneration beds exhaust or distillate columns begin to accumulate silica deposits that one truly assesses worst case.

For a sterilizer, this means demonstrating that heat and/or steam, radiation or gas can penetrate into the materials in the chamber to effect sterilization. Effective sterilization is usually demonstrated by challenging the chamber load with a known bioburden and then demonstrating kill of this bioburden.

Biological challenges of a sterilizer are traditionally made with spores. Because the spores of the bacteria *Bacillus* are highly resistant to destruction, they can provide a worst-case challenge. The level or amount of the biological challenge should be at least 10^6 spores per load. For overkill loads, a demonstrated 12 log reduction of bioburden is considered acceptable.[10–12]

Alternatively, for products or components that are sensitive to the effects of the sterilizing agent (heat, steam, gas, radiation), the sterilization cycle must be designed to expose the product to the sterilizing agent for the minimum amount of time while ensuring that the maximum amount of bioburden is killed. We do not discuss here the studies required to design these product specific cycles, but assume the use of an overkill sterilization cycle.

The PQ section of an equipment validation protocol describes preliminary operations, performance qualification procedures and acceptance criteria.

Preliminary operations include the validation of biological indicators such as spore strips or endotoxin,[11, 12] preparation of components to be sterilized and establishment of component loading configurations. All components used for validation work should be processed as they would be routinely to ensure that, for example, the moisture load, bioburden and particulate loads in the validation runs represent routine processing conditions. Loading configurations and maximum capacities must be established as well as cycle times and conditions.

In addition, production personnel should design sterilizer loads that represent loads used routinely during production activities. The loading of large numbers of identical components is straightforward. The loading of miscellaneous equipment—beakers, forceps, trays, spray bottles—may seem less controllable; it is not. Follow common practice in the wash area, and design loading configurations for miscellaneous equipment to be followed routinely.

The **worst-case challenge** is an often misunderstood concept. Clearly, the level of biological challenge—the number of spores in the load—can be considered in worst-case terms, especially when the bioburden of components prior to sterilization is routinely demonstrated to be far below the challenge level. The location of the biological indicators, however, is also important to the worst-case scenario—when spore strips are located in hard-to-reach places in the chamber, they are less likely to be killed. When autoclaving equipment with lengthy or convoluted silicone tubing, for example, you can create a worst-case challenge by placing a spore strip inside the tubing, far from either open end.

Because loading configurations directly affect the flow of sterilant throughout the chamber, they also must be considered in a worst-case context. It is natural to think, for example, that a full load of vials would be the worst-case condition for the hot-air sterilization of glass vials. In a facility that fills 2-ml, 5-ml, 10-ml, and 20-ml vials, for example, one might logically pick a full load of 20-ml vials as the worst-case challenge. It is not.

Sterilizing glass in a hot air oven requires bringing a large mass of glass to a set temperature and maintaining that temperature for a set length of time. As a result, the *mass* of the glass in the sterilizer—not the volume of space it occupies—most directly defines a worst-case situation. The worst-case load of vials would be a full load of 2-ml vials.

Over the years, however, the worst-case approach for some applications has developed into a minimum/maximum load approach. This is because the minimum load will often demonstrate the worst-case condition for some critical parameters.

When evaluating surface-to-liquid interactions in a liquid-fill vial product during terminal sterilization, for example, the minimum fill (2-ml) vial rather than the maximum fill (20-ml) vial often demonstrates worst-case conditions. This is because the surface area-to-volume ratio is greater for the 2-ml than for the 20-ml vial. Because it is difficult to predict all potential adverse effects, performing both minimum and maximum evaluations can be beneficial.

Performance qualification procedures. A performance qualification study should demonstrate that the equipment or utility system can consistently meet the operating specifications and perform its intended function (sterilization, depyrogenation, lyophilization) during at least three consecutive, successful cycles.

Before initiating a performance qualification, ensure that operating SOPs and processing forms accurately describe acceptable loading configurations and cycle parameters. Forms with a diagram of the empty chamber allow technicians to confirm, cycle-to-cycle, the location of materials and thermocouples in the chamber. Ensure that processing of materials used in validation studies is equivalent to routine processing. Then run three consecutive cycles for each configuration and collect all recording charts, processing forms and/or batch records for review.

Performance qualification acceptance criteria. Ensure that all processing parameters meet the specifications cited in the validation protocol, that the raw data are available to support these claims, that all biological indicators test sterile and that the number of consecutive runs meets requirements.

Validation Acceptance Criteria

When the IQ, OQ and PQ are complete, they must be reviewed as a unit, and at least one wrap-up operation must be performed—all thermocouples used in the study must be recalibrated to ensure that all data collected are accurate.

Equipment certification/validation certificate. When the validation is determined to be complete and acceptable, the validation committee issues a validation certificate to be posted on or near the equipment. This certificate should include the equipment number of the unit, the date validation is complete, reference to the validation protocol number and edition number and an expiration date for the validation. (Figure 9.1)

Categorization of Equipment/Utility Validations

The FQMP, described in the previous chapter, indicates that validations will be categorized. Use this categorization to assess the importance of each piece of equipment or utility in the facility according to its impact on the safety, efficacy and quality of final product.

Full validations (IQ, OQ, PQ) will be required for equipment and utilities that are unique to your operation, for equipment that is used in a manner other than that recommended by the manufacturer of the unit, or for critical processing or support equipment (sterilizers, lyophilizers, pure steam generators). Less rigorous evaluations may be appropriate for support equipment or utilities such as compressed air systems that *only* feed pneumatic controls. Less rigorous evaluations include IQ only, IQ and OQ, or IQ, OQ and an abbreviated PQ.

In determining the nature of an equipment or utility evaluation consider the following questions and concerns:

- Does the utility or the performance of the equipment directly impact the safety, efficacy or quality of final product?

Our Laboratories, Inc.   Page 1 of 1

Validation Certificate

The validation of: _____

Equipment ID# _____

was conducted according to Protocol, _____, Revision _____.

Validation was completed on _____.

This validation expires on _____.

_____   _____
Quality Control/Assurance   Date

_____   _____
Production   Date

_____   _____
Maintenance   Date

(Figure 9.1)

- What would be the impact of deviations or failure in these utilities or equipment?
- What is the likelihood that deviations or failures will occur?
- Can the utility/equipment be designed to prevent these failures? If not, is there any way to minimize the impact of failure?

Categorization requires good scientific sense and good business sense. It is used primarily to reduce the overall workload of validation while preserving the quality of the processing operations and the final product.

Validation Schedules

During the start-up of a facility and during annual revalidation activities, the order of equipment, utility and process validation is important. It usually follows that utility systems are validated prior to the initiation of equipment validation procedures, and all equipment and analytical methods are validated prior to the initiation of process validation procedures. Method and process validation concerns will be presented in the following chapters.

10

Master Method Validation Protocols

METHOD VALIDATION IS REQUIRED to determine if a given method is a suitable way to provide useful analytical data for a known set of samples. Therefore, analytical methods must be validated for specific uses. Revalidation may be required if instrumentation, sample composition or reagents change.

A master method validation protocol presents guidelines on how to develop method-specific validation protocols. It describes format, offers guidelines on content and provides a way to determine how much analysis is required for each assay system. It should offer a format for designing individual method validation protocols. A few good references provide supplemental reading for designing a method validation protocol.[13-16]

Once a master protocol is prepared, individual assay validation protocols can be written. The individual protocols provide detailed instructions about number of replicates, sample composition or dilution, equipment calibrations and other factors.

Consider the following sections when designing a master method validation protocol.

Method principles. This section describes the general principles at work in the analysis. Give a brief history of assay development, when appropriate. Cite any known sample requirements or assay sensitivities based on the principles of the assay.

Method suitability. Describe how the assay will be used, and when appropriate, why it is preferred to other methods.

Method categorization. Many analytical methods are performed in a quality control laboratory, and although all assays must be controlled, not all require a rigorous validation. As a result, methods should be categorized based on their importance in assessing the critical identity, strength, purity, safety and efficacy parameters of products, and/or the uniqueness of the principles or techniques used in the analysis.

If, for example, an assay system measures the potency of final product, it is appropriate to design a method validation protocol. If, on the other hand, an existing compendial method (heavy metals) or a generally accepted method (biuret protein analysis) is already in place, a less rigorous analysis may be sufficient.

Declare categories in the master method validation protocol and provide guidelines for their use. Three categories are usually sufficient—critical, non-critical and support, or Levels I, II and III as described in *USP XXII*, Chapter 1225.[15] The guidelines for the European Community offer more specific directives about which aspects of validation apply to what type of assay.[16]

Validation Procedures

Raw material control. The control of an assay system depends directly on the quality and on the control of raw material reagents. Review the reagents and components required to perform the assay. Ensure that part number specifications for these items adequately assess any critical parameters that would affect assay outcome. Review vendors for vendor-to-vendor inconsistencies. Change specifications as appropriate.

When you prepare reagents for assay systems, evaluate the stability of these reagents over time and in routine-use conditions. Provide guidance about adequate storage conditions and expiration dates for these reagents.

Sample acceptance criteria. Sample volume, dilution, buffer components, pH, temperature, osmolarity and turbidity can also directly affect the performance of an assay system. Indicate in the method validation protocol any sample-specific requirements. In addition, when appropriate, indicate a method for determining sample enhancement and inhibition of the testing method.

The master method validation protocol should also discuss any issues concerning the stability of samples. If samples are likely to change with time, provide guidance for assay performance within these time constraints.

Equipment control/calibration. Equipment quality, equipment calibration and maintenance methods also directly affect the assay procedure. Describe maintenance and calibration procedures for equipment used in the analysis. Refer to appropriate SOPs.

In some cases full validation of equipment may be required; in others, a routine calibration record is sufficient. When routine equipment cleaning, adjustment or calibration is required, events can be recorded chronologically in a log book, or the information can be formatted directly into the assay data collection form.

In the case of HPLC columns that are used for more than one type of sample analysis, a chronology of use, regeneration, cleaning and calibration must be kept. Procedures to evaluate column-to-column equivalency must be described and referenced.

Technician training. Technicians must be trained specifically to perform analytical assays, and this training should be documented. The technician training program can be designed in conjunction with the ruggedness section of assay validation. This will be discussed later in this chapter.

The method SOP. Before initiating the validation of a method, ensure that the assay SOP is available and that it accurately reflects the procedure as it will be performed. In addition, review the data collection form to ensure that all critical process control parameters will be recorded during validation procedures.

Method Evaluation

When designing a method validation protocol the following analytical performance characteristics should be considered: precision; accuracy; limits of detection; limits of quantitation; selectivity, specificity and interference; linearity and range; ruggedness; and correlation. Each final protocol should have its own set of method validation requirements based on an appropriate mix of the performance characteristics.

Precision is a measure of assay reproducibility. Measure (evaluate) the variation in homogeneous samples by performing several independent analyses and determining mean and relative standard deviation. A minimum of six replicates with a relative standard deviation (RSD) of not more than (NMT) 2% is acceptable for most methods.

Accuracy measures the compatibility of the assay value with a true or absolute value or standard. For example, demonstrate the ability of the assay to recover a known amount of analyte and express the result as a percentage.

Replicates of 6 and recoveries or accuracies of 60–110% are acceptable for concentrations below 100 ppb; recoveries of 80–100% are acceptable for concentrations above 100 ppb.

Limits of detection specify the lowest concentration of a sample that can be detected with the method (not necessarily the lowest amount that can be quantified).

When the assay is routinely performed test samples should contain 2–3 times the minimum amount capable of detection.

Limits of quantitation specify the lowest concentration of a sample that can be quantified with an acceptable degree of precision.

For instrument methods, determine the standard deviation (SD) of a series of blanks, and multiply the SD by 10 for an estimate of the limit of quantitation. Confirm this with samples prepared at that concentration.

Selectivity, specificity and interference are measures of the assay's sensitivity to impurities, related chemical compounds and degradation products of product excipients; they measure the degree of interference.

Spike known concentrations of sample with known concentrations of potential contaminants and impurities.

Linearity and range. Linearity demonstrates that the method can elicit results that are mathematically related to the concentration of the analyte in the sample.

Linearity is usually demonstrated over a defined range of analyte concentration.

The slope of a regression line and its variance provide a mathematical measure of linearity; the y-intercept is a measure of potential assay bias. Range is validated by demonstrating accuracy and precision at its extremes.

If the assay relationship is not linear (log/log or log/linear) demonstrate these relationships as well.

Ruggedness. The ruggedness of an assay is a measure of its reproducibility in the face of variations such as analysts, different instrumentation, lots of reagent or elapsed assay times.

Compare the reproducibility of the assay when challenged under extreme conditions with the precision of the assay under normal conditions. For example, to evaluate the ruggedness of the assay when performed by different technicians, have four analysts perform one assay per day for three days with identical samples and evaluate precision.

Correlation. If the assay replaces an existing assay or is used to support another assay which measures a directly related characteristic, it may be appropriate to evaluate the correlation between assays. Such an evaluation requires that identical samples be tested several times by both methods.

Routine Monitoring of Method

Each method validation procedure and each assay procedure should indicate routine monitoring requirements, as appropriate. These requirements may include blanks, positive controls, negative controls, reagent testing and/or instrument readings or calibrations.

These requirements allow for ongoing control of the procedure and ensure that the assay continues to meet validation criteria. Routine monitoring data can be used in audit activities for the method itself, for technician variability, for reagent change evaluations and for the change in seasons.

Revalidation requirements. The quality control or quality assurance department can initiate a revalidation of a method at any time. Significant changes in reagent vendors, instrumentation and technicians could trigger a revalidation.

Finally, when an individual method validation has been completed, initiate appropriate changes in assay procedures and data collection forms to support routine method process control. QA monitoring of trends in these assay parameters will help to assure that the method remains valid.

11

Process Validation Protocols

THE TERM VALIDATION WAS INTRODUCED into the vocabulary of Good Manufacturing Practices in 1976 as a part of the GMP revision adopted in 1978. This initial use of the validation concept applied specifically to sterilization. In 1987 the Food and Drug Administration issued two guideline documents that specifically defined process validation as follows: "Process validation is establishing documented evidence which provides a high degree of assurance that a specific process will consistently produce a product meeting its predetermined specifications and quality characteristics."[6,7]

These and subsequent documents describe process validation for specific aseptic manufacturing events. As the industry grows and changes, however, it is sometimes difficult to know when to apply process validation principles to specific aseptic manufacturing events.[17-20] All manufacturing requires the performance of routine operations. Some routine tasks are simple, some complex; some are performed by equipment, some are strictly manual; some are performed as a part of a more complex operation, some are performed in isolation. So how does one determine what types of activities are processes, and then how does one determine which ones might require validation?

Any final product sterilization event requires validation, but what type of validation? Product that is terminally sterilized in an autoclave, for example, can be validated as an extension of the autoclave equipment validation. In such a case product loading configurations must be established, and three runs of product (minimum and maximum loads) can be evaluated for sterility and product quality.

The process of filter-sterilizing a liquid product, however, can be more complicated.

Certainly the filter cartridge can be validated as a piece of equipment by demonstrating its ability to effectively remove bacterial contaminants from the product solution at established flow rates and pressures. But the routine use of the filter sterilization unit often involves extensive manual manipulation. For example, the filter housing may require sterilization and aseptic reassembly, and these critical manipulations could adversely affect the subsequent reliability and reproducibility of the filter sterilization event. It may be more appropriate, therefore, to consider filter sterilization of a liquid product as a critical processing event and validate it accordingly.

Candidates for process validation include all critical aseptic manipulations of final product; any product manipulation event whose failure could adversely affect the safety, efficacy or quality of final product; and any process result that cannot be adequately tested in the final product.

Biologic production, in particular, requires rigorous process validation. The molecules are highly susceptible to variable processing conditions. Because of their innate sensitivity to heat, they are often processed aseptically, and removal of bacterial, viral and protein contaminants cannot be adequately tested in final product.[21-23]

In order to determine the role of process validation in your production process, identify and evaluate all critical processing events. Each major processing step is a likely candidate when the reproducibility and reliability of the processing event is not strictly equipment-dependent, when the processed product is likely to be affected by variable process conditions, when aseptic techniques are involved, and when processing characteristics cannot be adequately evaluated in final product.

Protocol Format

The four sections of process validation protocol loosely follow the concepts of installation qualification (IQ), operational qualification (OQ) and process qualification (PQ) as presented for equipment validation protocols, that is, introduction, preliminary operations, process qualification and product qualification.

Introduction. The reason for process validation should be declared, because it can directly affect the processing and product criteria that will most effectively support the validation event. Process validation is required when the process is new or being changed. Production scale-up, new facilities, new production equipment, change in product components and change in formulation or dosage are all candidates for process validation.

When process validation results from change, it is instructive to describe how to evaluate the effects of that change—that is, how to define the key processing variables. Any impact on routine production should also be declared.

Preliminary operations. The preliminary operations section of the process validation protocol should list all activities that must occur before validation can begin. Consider raw material, equipment, utility, environmental, process and personnel requirements when designing this section of the process validation protocol.

When planning to validate an aseptic filling operation, for example, you must ensure the consistent quality of raw materials and their compatibility with one another. It is essential to ensure that component sterilizers have been validated, that all employees involved in the aseptic filling event have been trained in GMP operations, that the environment in which the processing will occur meets acceptance criteria, and that any minor processing events have been evaluated or validated. IQ, OQ and calibration requirements may also be required for any process-specific equipment.

List the requirements as events, methods of evaluation and acceptance criteria. When performing a process validation, document that the acceptance criteria—listed in the preliminary operations section of the process qualification—have been met before proceeding.

Process qualification. A process qualification is similar to an operational qualification for a piece of equipment; that is, it demonstrates that the process can proceed as described and meet equipment and process-dependent quality criteria. In aseptic processing events, for example, it is common to demonstrate that processing can occur with a minimal amount of contamination. Media fills are used to demonstrate this.

During a media fill, all equipment performs as it would for routine operation except that the product filled is microbiological medium. Because the medium will support the growth of any bacteria that it contacts during processing, it rigorously demonstrates aseptic technique. In addition, the media fill can effectively demonstrate the quality of other process operations—fill-volume controls on fillers, filling machine operation, stoppering operations, for example.

Media-fill vials can also be used to evaluate container/closure integrity. Store media-fill vials right-side up and then upside down for extended periods of time and observe them for growth. These evaluations can support any other studies that assess container/closure integrity and can be an important measure for evaluation when the process change involves a change in container or closure components.

The process qualification should be formatted as a batch record to facilitate the proper collection of data and samples. Numerous guidance documents discuss media fills for traditional aseptic processing events.[6, 18–20, 24, 28] Consult these references when designing a media challenge for a unique application.

When planning a process qualification, it is necessary to consider several factors. Which processing variables need to be evaluated in terms of upper and lower limits? How many media runs will demonstrate reproducibility? (Usually three are recommended.) How many units will be filled during each run? How long will the filling operation last? It should approach routine processing times. What challenges to the system will occur during the fill?

When planning a process qualification, also consider any challenges to the event. During a media fill in a cell culturing machine, for example, it may be appropriate to mimic sampling events, exchange of factor bottle, and/or exchange of gas cartridge events, etc. Challenge the unit or the process with all routine operations that might adversely affect system sterility.[25–29]

Declare the key processing variables to be evaluated during validation. Describe how they will be evaluated and what is considered an acceptable result.

Product qualification. After demonstrating that a process performs consistently and reliably without contaminating the product, the product itself must be processed.

In order to control the processing event, it is a good idea to write a formal production batch record for these product qualification events. This batch record describes the processing events step-by-step and provides space to document critical processing parameters. In addition, any special concerns or special sampling or monitoring requirements unique to the product qualification can be written into the batch record. The acceptance criteria for all processing parameters should be listed in the process validation protocol.

The product qualification event should be scientifically relevant and rigorous. Product qualification should be used to evaluate any processing procedures that might improve the flow of work, make it work or save time. For example, if it would be better to hold a sample overnight before completing QC testing, evaluate the effects of sample storage during this product qualification event. Similarly if it would save overtime or eliminate processing to reduce the mix time of a certain step, evaluate this effect during product qualification studies.

To complete a product qualification, it is common practice to run three identical product runs on identical equipment and produce three batches of product that meet all processing parameters as well as final product specifications. Product specification testing at the completion of product qualification studies can provide a preliminary assessment of processing quality. But it is the product stability studies which will ultimately support the changes evaluated during this qualification event.

Stability studies. Stability studies should be initiated with product qualification product to demonstrate the effects of temperature over long-term storage as well as any other factors that might cause product degradation such as light, oxygen, vibration and humidity. Usually the first GMP product batches that the FDA evaluates are used to support product expiration dating and labeling claims.

Worst-case scenario. A worst-case condition is often difficult to define for a process. Consider the length of the process event, the number of units processed, the processing environment, the number of individuals involved in processing and any routine manipulations that might compromise product sterility, potency, stability or quality.

A worst-case situation is not always the maximum load or fill. For example, when assessing the interactions of a liquid product with its container, the smallest container often has the greater liquid-to-surface-area ratio and, as a result, provides the greatest challenge for assessing leachables. Select worst-case scenarios carefully.

Validation Review and Approval

The validation committee must review the validation protocol and all original data for compliance. Any deviation from or failure of a processing event must be investigated and the reason for failure identified and resolved before another processing event can take place. Final approval of a process should be documented; a validation certificate similar to those issued for equipment is adequate.

When the process validation has been completed and preliminary results appear acceptable, review the process and select processing parameter limits for routine processing. Change appropriate SOPs, batch records and specifications to ensure routine testing or monitoring of these parameters so that they meet the acceptance range established during validation.

12

Good Business Practices

IF YOU ACCEPT THAT A QUALITY-BASED DOCUMENTATION SYSTEM is a corporate communications tool that facilitates interaction both inside and outside a company, then it becomes quite logical to extend such a system throughout a corporation, so that all corporate policies are documented.

To extend the fundamental concept of the GMP documentation system, consider a system for good development practices, good clinical practices, good purchasing practices, good marketing practices, good regulatory practices and good finance/accounting practices. Each documentation system would require descriptive documents that define and establish quality standards, and would require data collection documents to ensure standards are met routinely. Similarly, numbering systems and data files would facilitate the organization, review and retrieval of information.

Regulatory requirements dictating minimum documentation for each area of the corporation are set forth by agencies such as the Food and Drug Administration, the European Commission, the Securities and Exchange Commission, the Internal Revenue Service, the Occupational Safety and Health Administration, the Environmental Protection Agency and the International Standards Organization.

The regulator's minimum standards and policies would be supplemented with corporate standards and policies. Some of these are instituted to protect the corporation, the employees or the product; some are enacted in response to competition, and some support the ethical conduct of the company within the corporate community. Whatever the requirements or intent of these corporate commitments, they should be documented. Written declarations make clear to line employees and management within the corporation, as well as to outside reviewers, the rationale that supports consistent decision making throughout the corporation.

Be the Expert. Act Responsibly.

Designing a documentation system appropriate for every corporate application is not possible, and should not be expected. Each system must be designed to ensure the quality of unique operations and unique products, as well as to complement the cultural environment of the corporation.

A set of expectations is associated with the design and implementation of any quality-based system of control. You are the expert in your system, operation, cell line or product. Consequently, you are expected to know what matters and what does not matter—what could adversely affect the safety, efficacy or quality of your final product, and what is unlikely to affect it; you are willing and able to act responsibly with your expert knowledge; you are willing to adopt a GMP philosophy of quality assurance and adapt it to your project; you will use a combination of good scientific sense and good business sense to make the decisions necessary to establish and meet your quality standards and you will design and implement useful documentation systems, not systems designed solely to satisfy a perceived FDA requirement.

References

[1] Code of Federal Regulations, Title 21, Subpart J, Section 211.184 (U.S. Government Printing Office, Washington, D.C., 1990).

[2] Code of Federal Regulations, Title 21, Subpart E, Section 211.80d (U.S. Government Printing Office, Washington, D.C., 1990).

[3] Code of Federal Regulations, Title 21, Subpart E, Section 211.82, Part 211.84 (U.S. Government Printing Office, Washington, D.C., 1990).

[4] C.V. DeSain, *Drug, Device and Diagnostic Manufacturing: The Ultimate Resource Handbook*, 2nd edition (Interpharm Press, Inc., Buffalo Grove, Illinois, 1993).

[5] *Federal Register*, 41 (31), (13 February 1976), pp. 6878–6894.

[6] Center for Drugs Evaluation and Research, Division of Manufacturing and Product Quality, "Guideline on Sterile Drug Products Produced by Aseptic Processing" (Food and Drug Administration, Rockville, Maryland, 1987).

[7] Center for Drugs Evaluation and Research, Division of Manufacturing and Product Quality, "Guideline on General Principles of Process Validation" (Food and Drug Administration, Rockville, Maryland, 1987).

[8] "Biological Indicators," *USP XXII/NF XVII*, Chapter 1035 (U.S. Pharmacopeial Convention, Rockville, Maryland, 1990), p 1625.

[9] "Depyrogenation," *Technical Report #7* (Parenteral Drug Association, Philadelphia, 1985).

[10] "Validation of Dry Heat Processes Used for Sterilization and Depyrogenation," *Technical Report #3* (Parenteral Drug Association, Philadelphia, 1981).

[11] "Validation of Steam Sterilization Cycles," *Technical Report #1* (Parenteral Drug Association, Philadelphia, 1978).

[12] "Sterilization of Pharmaceuticals by Gamma Radiation," Technical Report #11, *Journal of Parenteral Science and Technology* (supplement) 42 (3S), 1988.

[13] J. Guerra, "Validation of Analytical Methods by FDA Laboratories I," (*Pharmaceutical Technology* 10 (3) pp. 74, 76, 78 (1986).

[14] M.J. Finkelson, "Validation of Analytical Methods by FDA Laboratories II," (*Pharmaceutical Technology* 10 (3) pp. 75, 78, 80–84 (1986).

[15] "Validation of Compendial Methods," U.S. *Pharmacopeia XXII*, Chapter 1225, p. 1710 (1990).

[16] Commission of the European Communities, "Analytical Validation," *Rules Governing Medicinal Production in the European Community*, Vol. 3 Addendum, (Office of Official Publications of the European Community, Brussels—Luxembourg, 1990), pp. 1–16.

[17] Center for Biologics Evaluation and Research, *Cytokine and Growth Factor Pre-pivotal Trial Information Package* (Food and Drug Administration, Rockville, Maryland, 1990).

[18] "Validation of Aseptic Filling for Solution Drug Products," *Technical Monograph #2* (Parenteral Drug Association, Philadelphia, 1987).

[19] "Validation of Aseptic Drug Powder Filling Processes," *Technical Report #6* (Parenteral Drug Association, Philadelphia, 1984).

[20] J. Wasynczuk, "Validation of Aseptic Filling Processes," *Pharmaceutical Technology* 10 (5), pp. 36–43 (1986).

[21] A.S. Lubiniecki, M.E. Wiebe, and S.E. Builder, "Process Validation for Cell Culture Derived Pharmaceutical Proteins," *Large-Scale Mammalian Cell Culture Technology*, A.S. Lubiniecki, Ed. (Marcel Dekker, Inc., New York, 1990), pp. 515–541.

[22] Commission of the European Communities, "Development Pharmaceutics and Process Validation," *Rules Governing Medicinal Production in the European Community*, Vol. 3, (Office of Official Publications of the European Community, Brussels—Luxembourg, 1989), pp. 3–10.

[23] Commission of the European Communities, "Validation of the Purification Procedures," *Rules Governing Medicinal Production in the European Community, Production and QC of Human Monoclonal Antibodies*, Vol. 3 Addendum, (Office of Official Publications of the European Community, Brussels—Luxembourg, 1990), p. 52.

[24] R. Tetzlaff, "FDA Regulatory Inspections of Aseptic Manufacturing Facilities," *Aseptic Pharmaceutical Manufacturing*, W.P. Olson and M.J. Groves, Eds. (Interpharm Press, Buffalo Grove, Illinois, 1987).

[25] Y.H. Chiu, "Validation of the Fermentation Process for the Production of Recombinant DNA Drugs," *Pharmaceutical Technology* 12 (6) pp. 132–138 (1988).

[26] M. Weiss, "Advantages and Validation Issues for Long Term Continuous Production," *Genetic Engineering News*, 10 (7) p. 8 (1990).

[27] M.J. Carter, and R.A. Scotland, "Validating Products Derived from Continuous Cell Lines," *BioPharm* 2 (10) pp. 24–27 (1989).

[28] "Microbial Evaluation and Classification of Clean Rooms," *Pharmacopeial Forum*, 17 (5) pp. 2399–2404.

[29] "Dispensing Practices for Sterile Drug Products Intended for Home Use," *Pharmacopeial Forum* 18 (2) pp. 3052–3075.